Food and the Mid-Level Farm

- move Ch. 1, to 1st or 2nd class lecture
    2

- look at Welsh, 1996 for consolidation
      (p. 55)   week?

- maybe don't need ch. 4?

Food, Health, and the Environment
Series Editor: Robert Gottlieb, Henry R. Luce Professor of Urban and Environmental Policy, Occidental College

Food and the Mid-Level Farm

Renewing an Agriculture of the Middle

edited by Thomas A. Lyson, G. W. Stevenson, and Rick Welsh

The MIT Press
Cambridge, Massachusetts
London, England

© 2008 Massachusetts Institute of Technology

For information about special quantity discounts, please email ⟨special_sales@ mitpress.mit.edu⟩

This book was set in Sabon on 3B2 by Asco Typesetters, Hong Kong. Printed on recycled paper and bound in the United States of America.

Library of Congress Cataloging-in-Publication Data

Food and the mid-level farm : renewing an agriculture of the middle / edited by Thomas A. Lyson, G. W. Stevenson, and Rick Welsh.
   p.   cm.—(Food, health, and the environment)
Includes bibliographical references and index.
ISBN 978-0-262-12299-3 (hardcover : alk. paper)—ISBN 978-0-262-62215-8 (pbk. : alk. paper)
1. Family farms—United States. 2. Farm produce—United States—Marketing. I. Lyson, Thomas A. II. Stevenson, G. W. III. Welsh, Rick. IV. Series.
HD1476.U6F66 2008
338.10973—dc22   2007031164

10 9 8 7 6 5 4 3 2 1

To Fred Buttel and Tom Lyson, friends, colleagues, and scholars

Contents

Series Foreword

I am pleased to present the third book in the Food, Health, and the Environment series. This series explores the global and local dimensions of food systems, and examines issues of access, justice, and environmental and community well-being. It includes books that focus on the way food is grown, processed, manufactured, distributed, sold, and consumed. Among the matters addressed are what foods are available to communities and individuals, how those foods are obtained, and what health and environmental factors are embedded in food-system choices and outcomes. The series focuses not only on food security and well-being but also on regional, state, national, and international policy decisions as well as economic and cultural forces. Food, Health, and the Environment books provide a window into the public debates, theoretical considerations, and multidisciplinary perspectives that have made food systems along with their connections to health and the environment important subjects of study.

Robert Gottlieb, Occidental College
Series editor

Preface

The structure of agriculture in the United States is moving toward two relatively separate spheres: large, corporately coordinated, agricultural commodity production units; and dispersed, local, and smaller-scale farms relying on direct markets. The result is that midsize family-operated farming operations are finding it increasingly difficult to identify appropriate market outlets for their products. That is, midsize farms tend to produce volumes too large for direct markets, and cannot compete economically against the coordinated and corporate-dominated commodity systems that articulate with national and international marketing and distribution systems. In this sense, midsize farms and ranches are being squeezed out of agriculture through a mismatch with available markets. As Amy Guptill and Rick Welsh point out in this volume, "Rural sociologists and agricultural economists have traced the cost-price squeeze undermining U.S. agriculture to a production-marketing-processing system that maximizes the profit accruing to large agricultural corporations at the expense of producers. This system is increasingly dependent on contracting to maintain vertically integrated production chains."

If current trends continue, the structure of U.S. agriculture will encompass a small number of immense corporate-linked or owned farms, and large numbers of small direct-market operations. The former system of production will produce the vast majority of the food most Americans will consume. It will do so by producing uniform raw materials that are differentiated in the processing and especially marketing stages. The direct-marketing sector composed of small to extremely small farms will take a different marketing approach with differentiation taking place at the point of production. These farms will produce small volumes of many different species and varieties of animal and plant products with minimal processing and branding. Income for the vast majority of members of the small farm sector will originate from off-farm sources. In fact, the latest census of agriculture data shows that 90 percent of the income accruing to farms with less than $250,000 in annual sales (98 percent of all farms) comes from off the farm.

It is our contention that such a starkly bipolar production and marketing structure will result in a hollowing out of many parts of the rural United States, and will have important consequences for the future social viability of agriculturally supported rural communities. Critical issues such as the maintenance of municipal tax bases, job creation, population growth, environmental quality, and ultimately rural community welfare are at stake. The decline of the agriculture of the middle will hit regions such as the Midwest particularly hard.

The most likely solutions to this problem are threefold: scale up the aspects of direct markets that attract consumers, such as knowing who produces one's food and how particular farming systems protect the environment; develop organizations and strategies for empowering midsize farmers and ranchers to access developing markets for high-quality food products differentiated by dimensions such as ownership arrangements, sustainable production system, regional location, and "fair" business relationships within a supply chain; and develop efficient and targeted public policies that directly address the challenges and opportunities associated with the agriculture of the middle. The chapters in this volume look at each of these issues.

Food and the Mid-Level Farm: Renewing an Agriculture of the Middle is designed to serve as an intellectual and conceptual framework for moving private business development, public policy, and other supportive activity forward to address agriculture of the middle issues. Toward that end, we have involved many important analysts and authors from a variety of disciplines including sociology, political science, and economics who have a long and productive history of working on farm-level food distribution and consumer issues.

This volume is part of a larger project that was inspired by research and outreach activities begun at the Leopold Center at Iowa State University in the 1990s. In the early 2000s, funding from the Kellogg Foundation's Food and Society initiative and the U.S. Department of

Agriculture's Sustainable Agriculture Research and Education Program enabled the Agriculture of the Middle project to go national in scope. As described on the Agriculture of the Middle Web site (⟨http://www .agofthemiddle.org⟩), the project "seeks to renew what is being called the agriculture of the middle. This term refers to a disappearing sector of mid-scale farms and related agrifood enterprises that are increasingly unable to successfully market bulk agricultural commodities or sell products directly to consumers."

definition

Narratives of Renewal

Participants in the national Agriculture of the Middle initiative are composed of activists, farmers, food industry leaders, government officials, and scholars representing a diversity of regional, cultural, organizational, and food system standpoints. For example, besides the editors of this volume and other academics, task force leadership includes farmer leaders such as Adrian Wadsworth of Maine, Stephen Walker of Arkansas, Bob Quinn of Montana, and Keith Mudd of Missouri. Other participants include Theresa Marquez from Organic Valley, a "new generation" cooperative based in Wisconsin.

A critical aspect of this collective effort is to work toward establishing organizations and relationships committed to, and capable of, changing the food system so that it works for midsize farms. Yet in order to accomplish this goal over the long term, the food system must also work for nonfarm actors within the food system such that potentially all can benefit and prosper from the production, processing, distribution, and consumption of food and other agricultural products. As Robert Handfield and E. Nichols Jr. put it in *Supply Chain Reaction* (see chapter 7, this volume), "The only solution that is stable over the long term is one in which every element in the supply chain, from raw material to end consumer, profits from the business. It is shortsighted for businesses to believe they can solve their ... problems by punishing suppliers and customers."

A number of innovative organizations have emerged to serve as models for the restructuring of our food production and manufacturing system in order to address these issues. Two such organizations are Organic Valley/Cooperative Regions of Organic Producer Pools (CROPP) and the Organic Farmers Agency for Relationship Marketing (OFARM; see chapter 4, this volume). Organic Valley/CROPP has an established track record of maintaining organic price premiums for family-owned organic farmers, paying interest to farmers on their equity investment, a shorter than average return on their equity investment, and protecting the environment through the promotion and expansion of organic farming practices. And OFARM was formed specifically to address the "conventionalization" of the organic industry in order to maintain a space for family farmers earning their livelihoods from farming. As two OFARM members explain in chapter 4,

On the conventional side we've seen what's happened—low prices for farmers and a market controlled by the bigger buyers, like Cargill and the rest.

We want a fair and decent price for the grower, consistency in the market. We want to unite the organic community more than it is, so that they don't become a product of the conventional market. We don't want to conventionalize the organic market. We don't want to let it be taken over by the big conglomerates where we're slave labor out here. We want to maintain smaller farms.

Examples of nonorganic farmer-led organizations trying to maintain and expand a place in the food system for mid-level farms include Oregon Country Natural Beef. Its mission statement illustrates the need to create organizations committed to principles of farmer control, environmental protection, and more direct links with other parts of the food system. For example, "The co-op will be market driven and producer controlled from start to finish. It will make use of a livestock pool from diverse geographic areas, utilizing cattle that fit their environment, and sorting procedures that allow for more consistency on the rail. A consistent, high-quality product with expanded name brand recognition will respond to changing customer demand, and new product development will continue."[1]

Another example is Yakima Chief Inc., a hop producing, processing, warehousing, and marketing company located in Washington State. Yakima Chief is interesting for its commitment to serving the brewing industry through a farmer-controlled organization and because it controls a significant percentage of the total hop acreage in the United States. The company exemplifies both the need and possibility of moving beyond niche marketing to establishing organizations that serve markets in a sustainable manner but on a large scale.

In addition to these innovative producer organizations, <u>alternative re-tail outlets are</u> being established that serve midsize producers. As G. W. Stevenson and Rich Pirog argue in this volume, groceries such as New Seasons Market in Portland, Oregon, operate to serve the diverse needs of "reflexive consumers" seeking satisfying, healthful, and responsibly produced food products. But its <u>products "are sourced through agree-ments with regional producers where the price is based on what</u> the pro-<u>ducers need and what New Seasons Market can afford.</u>"

The connecting thread of this volume is that creating an equitable, environmentally benign, prosperous, and satisfying food system will re-quire purposeful policies, collective efforts, and cooperation of different types as well as among different social actors. It is fundamental to subor-dinate shorter-term self-interested strategies to longer-term shared goals and publicly oriented outcomes.

Plan of the Book

In part I, "Agriculture of the Middle: Why Farm Structure Matters," Fred Kirschenmann and colleagues explain the current trends in the structure of U.S. agriculture, and make the case for renewing the agricul-ture of the middle. Mike Duffy provides a companion piece (see the appendix) in which he uses data from the U.S. Census of Agriculture to document changes in the Agriculture of the Middle across a range of de-mographic, structural, and geographic categories. Kenneth A. Dahlberg analyzes the vulnerabilities of the concentrated pole of the emerging U.S. agricultural and food system.

Part II, "Organizational Structures That Could Support an Agriculture of the Middle," makes the case for the role of <u>collective bargaining</u> and other collective measures such as coordinating marketing among organic midsize grain farms to enable midsize farms to better articulate with the corporate-dominated mass markets. The increases in contract production and its meaning for midsize agriculture are also analyzed. A key compo-nent of the arguments in this part is that <u>farmer organizing,</u> trust, and <u>the realization of collective interest is critical</u> for realizing a resurgence <u>of the agriculture of the</u> middle. For example, Thomas W. Gray and Ste-venson posit that agricultural cooperatives, because they are democrati-cally oriented business entities, can be an "organizational and mobilizing

need for farmer organizing

instrument for realizing some of the agriculture of the middle goals." Likewise, Guptill and Welsh assert that OFARM's goal is not just to capture more value in the food chain but rather to participate in the governance of the industry; "and its key strength lies in the trust among participants that enables the flow of information [between them]." In this way, the negative outcomes attributed to contract production and coordination in agriculture by Mary Hendrickson and colleagues hopefully can be avoided.

In part III, "Bringing Mid-Level Supply Chains and Consumers Together," the authors take on the task of envisioning how growing markets for significant volumes of high-quality, differentiated food products can provide a platform for regenerating an agriculture of the middle. Stevenson and Pirog provide the outlines for new, values-based food supply chains in which farmers and ranchers are treated as strategic partners, rather than interchangeable (and exploitable) input suppliers. In such value chains, farmer/rancher identities and brands are preserved, and business agreements are experienced as fair and for appropriate time frames. As Eileen Brady and Caitlin O'Brady point out in their chapter,

The consumers of this emerging market want to know more about where their food comes from. They want assurances that their food choices are healthy and safe. They want the story behind the product. They want accountability. Who produced the food they are eating? How was it grown? What's the name of the farmer? The new farmer, rancher, or fisher must realize they are selling more than the product they are producing. They are building a new relationship—a relationship that requires participation and interest from the producer as well as the consumer. Producers must reach out and understand their consumers.

In the final part, "Policies That Could Support an Agriculture of the Middle," a number of policy alternatives are critically examined. Daryll E. Ray and Harwood D. Schaffer contend that to preserve the middle of agriculture, we need to devise, among other interventions, an international supply management and price support program for the major crops. Thomas A. Lyson takes a different tact and offers an array of diverse approaches for embedding agricultural operations in rural communities to create a civic agriculture that nourishes small to moderate farms. And Shelly Grow and colleagues show that state laws can enable the formation of collective bargaining cooperatives, which can bring with them important marketing services as well as the ability to balance the power of processing firms in regard to the fairness of contracts. For her part, while Sandra S. Batie agrees that green payment programs such as the Conservation Security Program (CSP) potentially offer benefits to midsize producers through payments for environmental services, she argues that such payments are not an efficient means to influence structural dimensions within agriculture. In a parallel fashion, Peter Carstensen examines the role and limits of antitrust policies to affect the types of changes needed to preserve midsize farms.

Given the renewed interest in preserving and strengthening an agriculture of the middle in the United States, the subject matter covered and the analytic frameworks employed are new and significant. Current volumes on agriculture, farm structure, and the food system have yet to address agriculture of the middle issues in any systematic fashion. We also believe that there is a tendency for many involved in topics related to the structure and sustainability of U.S. agriculture to accept the demise of midsize farms and ranches as either the natural functioning of economic markets, or a policy battle that cannot be won. We obviously disagree with such perspectives and resignation. In our view, it is time to refocus the attention of consumers and other participants in the food system on the midsize food production sector that has undergirded most of what we have valued about U.S. farming. Doing this can align powerful dynamics capable of altering the current trajectory within U.S. agriculture to the benefit of rural areas as well as consumers and citizens everywhere.

Note

1. Available at ⟨http://www.oregoncountrybeef.com⟩ (accessed January 8, 2008).

I

Agriculture of the Middle: Why Farm Structure Matters

1

Why Worry about the Agriculture of the Middle?

Fred Kirschenmann, G. W. Stevenson, Frederick Buttel, Thomas A. Lyson, and Mike Duffy

During the past several decades, the U.S. food system has increasingly followed two new structural paths. On the one hand, small-scale farm and food enterprises in many regions have thrived by adapting to successful direct markets that enabled them to sell their production directly to consumers. This is an encouraging trend with real benefits to the local communities. On the other hand, giant consolidated food and fiber firms have established supply chains that move bulk commodities around the globe largely to serve their own business interests.[1]

This new pattern of food systems has had a disastrous effect on independent family farmers—it has led to a disappearing agriculture of the middle. These farms and enterprises of the middle have traditionally constituted the heart of U.S. agriculture. They operate in the space between the vertically integrated commodity markets and the direct markets. While the bulk of these farms have gross annual sales of between $100,000 and $250,000, it would be a mistake to characterize them simply as midsize or small.[2] Many of these endangered agriculture of the middle farms are what the U.S. Department of Agriculture's (USDA) Economic Research Service calls "farming-occupation farms" and "large family farms."[3]

What we are calling the agriculture of the middle is, in other words, a market-structure phenomenon. Strictly speaking, it is not a scale phenomenon. Yet while it is not scale determined, it is scale related. That is, farms of any size may be part of the market that falls between the vertically integrated, commodity markets and the direct markets. But the midsize farms are the most vulnerable in today's polarized markets, since they are too small to compete in the highly consolidated commodity markets, and too large and commoditized to sell in the direct markets.

Ironically, it is also the midsize farms that have a comparative advantage in producing unique, highly differentiated products. Their smaller size enables them to remain flexible and innovative enough to respond to highly differentiated markets. And currently the demand for such products is increasing dramatically, especially in the food service industry. These products are suitable for the market of the middle. The commodity markets are ill equipped to produce such unique, highly differentiated products, owing to the uniformity and specialization demanded of commodity markets. And the direct markets are unlikely to produce the quantity of unique products that this emerging market demands. Furthermore, direct marketing will only affect the management of a fraction of our agricultural lands. As Patrick Martins, director of Slow Food USA, puts it, "Community supported agriculture programs, wonderful as they are, can't by themselves save American agriculture."[4]

This situation presents us with a rare market opportunity. There is a burgeoning market demand for foods that are produced in accordance with sustainable agriculture standards, and it is precisely the farmers of the middle who are in the best position to produce those products. What is missing is a functional value chain to connect these farmers to the markets. Our main thrust will be to help these farms develop competitive alternatives to commodity agriculture—alternatives that can potentially be much more sustainable economically, socially, and environmentally.

Nationally, midsize farms still make up the largest share of working farms—farms where the chief source of income and primary occupation is farming. These farms also constitute the largest use of farmland and currently remain a critical variable in rural community success. But the polarizing forces in the present market climate are rapidly driving these farms out of business.

These polarizing forces threaten to hollow out many regions of the rural United States by transferring most of the agricultural economic activities that have sustained rural communities, thereby impacting agribusiness viability, job creation, and the maintenance of local tax bases. And because these are mainly farms that have been in the family for several generations (and good land stewardship is a high priority since they regard their land as part of the family's heritage and local ecological knowledge has been handed down from one generation to the next), they make important social and environmental contributions.

While the majority of farmland in the United States is still managed by farmers whose operations fall between the two marketing extremes, if present trends continue, these farms, together with the social and environmental benefits they provide, will likely disappear in the next decade. The public good that these farms have supplied in the form of land stewardship and community social capital will disappear with them.

The phenomenon of the disappearing middle, of course, did not emerge in a vacuum. Changes in the structure of agriculture that helped to bring about the disappearance of the middle have been occurring for some time.

How Do Declining Farm Numbers Change U.S. Agriculture?

↓ population

Farm populations in the United States have been declining for more than half a century. In fact, by the early 1990s Calvin Beale at the USDA had begun to refer to the steady decline in farm populations between 1950 and 1980 as a "free fall" situation leading us toward "trauma."[5] Certainly, many people in the United States continue to enjoy a surplus of food and fiber despite these steady declines in farm populations. And many agricultural experts continue to see the attrition of farmers as a necessary "market correction," insisting that depressed farm economies are due to inefficiencies. In their minds, we still have "too many farmers."

So are declining farm populations leading to trauma or maximum efficiency? If fewer farmers are able to produce more than enough food and fiber to meet our domestic and export needs, why should we worry about declining farm numbers at all? Many policymakers and perhaps the general public are, in actuality, not concerned. At a meeting that took place at the National Academy of Sciences Board on Agriculture almost twenty years ago, an official of the Office of Management and Budget remarked that "if two or three farmers can produce all of the food and fiber we need, who cares? In fact, if robots can do it, who cares?"[6]

But farm numbers are not the only issue at stake. If we are simply asking our farmers to produce bulk commodities to be manufactured into

It's not the issue

food, fiber, energy, and other products as cheaply as possible, without regard for the social and ecological costs associated with such production, then we might indeed want to stay the present course and reduce farm populations to the lowest possible number. But we have traditionally expected more from our farmers. We expect them to take care of the land for future generations, care for their animals properly, and protect the environment. We expect them to be good citizens of their communities. We want them to supply us with food products that have unique attributes. And we rely on them to provide us with food security. All of these public aspects contribute to a healthy landscape, healthy communities, pleasurable eating—and a sustainable future.

The USDA's *A Time to Choose: A Summary Report on the Structure of Agriculture*, published in 1981, pointed to some of the crucial issues facing agriculture that touch on these expectations. The report warned that the structure of agriculture that we choose will "shape the options available for generations to come and...affect the quality of life of all citizens." It went on to suggest that it was time to "make choices" concerning "our immediate needs" and "the needs of future generations," between "the maximization of current production and exports and long-run resource utilization and conservation." The "most critical" choice of all, the report observed, was "deciding what structure of agriculture" could meet those goals. The report also suggested that "there can be little doubt that one of the most important tasks before us is maintaining the productive capability of our resource base over the long term," and that "the market may fail to adequately reflect the full costs of resource use over the long run."[7] Nothing has happened in the more than twenty-five years since to alter that assessment. Everything that has happened makes that call to action more urgent than ever.

The central question still facing us is whether we can reasonably expect farmers to provide these public services within the framework of the current structure of the food and agriculture system we have developed. We have now reached a crossroad. This is not just about farm numbers or "saving the family farm." The decline in farm populations, as the USDA report noted, is closely linked to the structural changes that drive that decline, and the disappearing middle plays a key role in that decline. Consequently, as we enter the twenty-first century, a whole segment of the food and farming industry—the agriculture of the middle—is about to become extinct. And the reason we are calling attention to this development is that it will dramatically change the very landscape of the rural United States, jeopardize the future productive capacity of the land, and severely limit our food choices.

The Disappearing Middle

Why a concern

Evidence of the disappearing middle is already accumulating. Iowa serves as a compelling example. The decade from 1987 to 1997 saw an 18 percent sales increase in farms that are 1 to 100 acres in size, and a 71 percent sales increase in farms that are more than 1,000 acres in size. Farms that range in size from 260 to 500 acres averaged a 29 percent decrease in sales. The percentage of operators and acres in all farms between 100 and 999 acres in size declined 23 and 25 percent, respectively.

In the time since the USDA's 1997 data were published, we have seen the middle disappear at an even more alarming rate. In Iowa during the five-year period from 1997 to 2002, there was a 17 percent drop in farms with sales ranging from $5,000 and $500,000, while the number of farms with gross sales of more than $500,000 increased by 17 percent. Farms with less than $2,500 of gross sales increased by 39 percent.

Of course, statistics vary from region to region. Many southern states have seen a drop in farms with gross sales of more than $500,000, while California, Washington, Alabama, Hawaii, and Rhode Island have witnessed a drop in farms with gross sales under $2,500. But the pattern repeats itself often enough to demonstrate that the bipolarization of the food system into direct and commodity markets is scale related.

Following are some additional examples of the changes in farm numbers that have taken place in the percentage of total sales during the five-year period between 1997 and 2002. While the percentages vary from state to state, owing largely to differences in the value of the commodities being produced as well as differences in climate and rainfall, the general pattern of the disappearing middle seems clear.

The impact of these trends is further exacerbated by the fact that the age distribution of farmers has altered dramatically in the past twenty-five years. In Iowa, for example, there were almost three times as many farmers under age thirty-five as over age sixty-five in 1982. By 1997, those numbers had exactly reversed. The national age distribution data appear to mimic those in Iowa. These disturbing statistics indicate that we do not have much time left to make significant changes in our food and farming system. Once the independent farmers and their diversified farms are gone, we will have lost the human capital necessary to make the changes we need to meet our goals of providing our citizens with the food choices they want, while restoring the ecological health of the land and revitalizing our local communities.

So, exactly what is it that we stand to lose if the agriculture of the middle disappears?

- The opportunity to choose foods with special desirable attributes
- Open spaces that are easily accessible
- Wildlife habitat
- Clean air
- Soils that hold rainwater for aquifers
- Soils in cropland and pastureland that help reduce flooding
- Taxes will increase because farmland requires fewer services than residential areas
- Diversified farmland that includes perennials serves as a carbon "sink" to reduce greenhouse gases that are implicated in global climate change
- A shift from smaller farms on a diverse landscape to endless fields of monocrops

Changes on the Landscape

The emerging bipolar food and agriculture system increasingly will dictate how farm management decisions are made. While direct-market farms generally are able to operate independently, and make land management choices that benefit the social and ecological communities in which they exist, they can never manage significant amounts of land. When a farmer grows, prepares, processes, and delivers a food product, there is a severe limit on the amount of acres they can manage. Meanwhile, farmers linked to large consolidated firms through contract relationships will be less able to make independent management decisions. They must make choices that serve the business interests of the consolidated firms with which they contract, and those decisions may not always benefit the community in which the farm exists.

Rapid consolidation, initially in the seed and manufacturing sectors, but now in the huge food retail sector, means that in the near future about six multinational retail firms will determine not only the size of U.S. farms but the type of management decisions made on them.[8] This has serious, long-term implications not only for U.S. food security but also for food production and processing enterprises.

Independent farmers, selling their production in the free market, have always made on-farm decisions based on a variety of intended outcomes. In addition to managing the farm for profitability, most farmers also made decisions that assured the survival of the farm in its particular community so that it could be passed on to future generations in good health. This is not to suggest that small, independent farms have always been managed to prevent soil loss, protect water quality, or maintain vibrant communities. There is a long history of degradation and loss that belies such a romantic picture of the yeoman farmer of the past.[9] But it is to say that many independent farmers have included these nobler considerations in their management decisions as a way of ensuring the health of the farm for future generations. As farmers increasingly enter into contractual arrangements with highly consolidated firms, such considerations will be ignored.

The commercial interests that drive these large consolidated firms are based on three primary business objectives: the development of supply chains, biological manufacturing, and the reduction of transaction costs.[10] Each of these business objectives will have a profound effect on how local farms are managed.

The development of supply chains means that on-farm decisions will no longer be made to benefit the long-term sustainability of the farm, the good of the community, or the health of the natural resources that sustain the farm. Decisions throughout the supply chain will be made solely on a profit basis to help a given large enterprise compete effectively with other supply chains to gain a larger share of the consumer's food expenditures.

The introduction of the concept of biological manufacturing means that farmers can no longer produce commodities based on what is best for the normal functions of the animals on the farm, the diversity of the landscape, or the general health of the farm. Rather, farm management necessarily will be focused on technologies designed to produce uniform products that meet the desired processing and retail objectives of the firm.

And the need to reduce transaction costs means that consolidated firms will do business only with the largest farmers. It simply is less costly to contract with one farmer who raises ten thousand hogs than it is to issue contracts to ten farmers who each raise one thousand hogs. All but the largest farms will become "residual suppliers."

The combination and broad scale impact of these three objectives may well lead to huge losses of both biological and social diversity, increasing standardization, and the simplification of complex natural and social systems.

Social and Economic Transformations

Given that farms will be pushed to these new levels of specialization, concentration, and homogeneity, a new generation of profound changes will take place on the landscape. First, farms will be replaced by industrial centers. In Iowa, for example, some have suggested that the farms of the future will consist of 225,000-acre industrial complexes. This transformation would reduce drastically the number of "farms" in Iowa.

Some argue it will be necessary to consolidate farms into such industrial assemblages to gain access to markets and negotiate effective prices with input suppliers. Surely such farms will not buy equipment from local dealers or fertilizer from local suppliers, nor will they deliver grain to local elevators. As with other industrial complexes, labor will consist largely of minimum wage earners.

For the most part, commodities produced on farms will be owned by the consolidated firms that issue the contracts. Just as Tyson retains the ownership of chickens placed on farms to be raised for them, using their feed and managed in accordance with their management plan, so other livestock species and patented seed crops increasingly will be owned by and raised for firms in accordance with the firms' management plans, using only the firms' technology and inputs. In 1992, *Time* magazine had already begun to refer to the contract "farmers" raising Tyson's chickens as "serfs on their own land."[11]

$ leaves community

This is a future, in other words, where all local business transactions will be made with distant supply chains, the benefit accruing not to local rural communities but likely to shareholders who live in far-off places. And the farmers who provide the labor for these operations will be allowed only minimal independent judgment and creativity. Like any other franchised business, such franchised farms will be given the freedom to operate the farms in accordance with the firms' directions and to accept most of the liability. In effect, this further transformation will amount to emptying the rural landscape of all its local agriculturally related economies and local talent. And it will deprive our food and agriculture system of the innovation and entrepreneurship that have been part and parcel of the independent owner/operator farm.

Amid today's greatly magnified concerns about homeland security, we should be aware that the changes in the nation's food system are harming our ability to control our own food supply. Food produced on large, far-flung farms is not likely to be reliably available in times of disaster, whether natural or human-made. Food that originates locally from a multitude of midsize farms faces far less risk from terrorist attacks. As David Orr points out, "A society fed by a few megafarms is far more vulnerable to many kinds of disruption than one with many smaller and widely dispersed farms.... [N]o society that relies on distant sources of food, energy, and materials or heroic feats of technology can be secured indefinitely."[12]

Biophysical Transformations

In addition to such social and economic transformations, there will be a biophysical one on the landscape. We know from past experience that large industrial complexes, owned by absentee landlords and managed by a highly centralized managerial class, do not exhibit a commitment to the care of the environment in which they exist—witness Love Canal, Louisiana's "cancer alley," the burning Cuyahoga River, the PCBs in the Hudson River, the dried-up Rio Grande River, and the hazardous waste inserted into farm fertilizer in Quincy, Washington. There is no good reason to believe that industrial farm complexes will operate with any higher degree of environmental care than any other industrial

No care of environment

complex—indeed, some "factory farm" poultry and hog complexes already serve as harbingers of the future of industrial agriculture. When short-term economic returns are the only business motivation of a firm, then long-term ecological, social, and human health will inevitably be compromised. Farms are no exception.

Specialization being one of the means to achieve efficiency in industrial operations, each of these agricultural industrial complexes likely will specialize in the production of only one or two commodities. That will foster additional biophysical degradation. We now know that imposing specialization on any ecosystem causes a host of ecological problems. These problems include the elimination of the biodiversity that is essential to the resilience and productivity of any ecosystem.[13] Furthermore, the uniformity and specialization demanded by this new level of industrialization invites genetic uniformity, which in turn leads to additional vulnerability.

Again, the poultry industry serves as a portent of the future. William Heffernan reports, for instance, that "90 percent of all commercially produced turkeys in the world come from three breeding flocks."[14] Such genetic uniformity, initiated to obtain a standardized product, results in birds with such compromised immune systems that their health cannot be maintained without the extensive use of antibiotics.

Farms, of course, are ultimately micro- (or small) ecosystems that exist within macro- (or larger, more complex) ecosystems. Every farm is an inevitable part of the larger dance of life—part of that complex, interdependent web of life that evolved (and continues to evolve) over almost four billion years. We ignore that evolving complexity at our peril.

The standard industrial answer to this cautionary tale is that we will always have the technological capability to restore any damage we may do to the ecosystem—especially with the newly discovered technological capacity of genetic engineering. We now seem to have convinced ourselves that we can redesign life to live better in a new biological order of our own making with our technological prowess.

In his book *The Future of Life*, E. O. Wilson analyzes the technological optimism that believes that we can redesign nature with technology to maintain its vitality, and he gives the proper response to such misplaced optimism.

Such is the extrapolation end point of technomania applied to the natural world. The compelling response, in our opinion, is that to travel even partway there would be a dangerous gamble, a single throw of the dice with the future of life on the table. To revive or synthesize the thousands of species needed—probably millions when the still largely unknown microorganisms have been cataloged—and put them together in functioning ecosystems is beyond even the theoretical imagination of existing science. Each species is adapted to particular physical and chemical environments within the habitat. Each species has evolved to fit together with certain other species in ways biologists are only just beginning to understand.[15]

Wilson is speaking here of whole ecosystems not farms. But again, farms are simply micro-ecosystems within macro-ecosystems. Consequently, anywhere agriculture is practiced, it must become part and parcel of the task of restoring our natural capital by restoring the species richness that is as essential to a healthy farm as it is to a healthy ecosystem. From this perspective, industrial agriculture with its specialization, centralization, and uniformity is simply another example of what Wilson calls "mistaken capital investment."[16] We must now redesign agriculture so that it becomes an integral part of restoring the landscape's biodiversity.

And the reason that the human resource factor on farms is important to that task is that such restoration is not likely to be accomplished without caring people on the land. As Wendell Berry reminds us, "There is a limit beyond which machines and chemicals cannot replace people; there is a limit beyond which mechanical or economic efficiency cannot replace care."[17]

Technological Transformations

A third likely effect from this new level of industrialization is the further promotion of authoritarian technologies. Lewis Mumford, arguably one of most important social critics in the United States, pointed out that from Neolithic times to the present, two technologies have "recurrently existed side by side"—one authoritarian, and the other democratic. Authoritarian technology, while powerful, is "inherently unstable." Democratic technology, while relatively weak, is "resourceful and durable." Mumford asserts that democratic technologies usually consist of "the small-scale method of production, resting mainly on human skill...remaining under the active direction of the craftsman or farmer, each group developing its own gifts, through appropriate arts and social ceremonies, as well as making discreet use of wide diffusion and its modest demands...[and has] great powers of adaptation and recuperation."[18]

Authoritarian technologies, conversely, tend to be large scale and concentrate power in the hands of the few. They rest mainly on high-tech inventions and scientific discoveries. They are generally under the direction of centralized management, usually exploiting the gifts of nature to suit the purposes of management. Because of its centralization and insatiable demands, authoritarian technology has little power of adaptation or recuperation.

These are some of the losses we will experience with the disappearance of the agriculture of the middle. R. Edward Grumbine reminds us that we can provide sound ecological management for natural systems only if we have someone living in those systems long and intimately enough to learn how to manage them.[19] Technology will not be a substitute for such wisdom. And since (as noted earlier) farms are micro-ecosystems within macro-ecosystems, the same holds true for farm management. It is precisely the farmers in the agriculture of the middle who fit that description and currently manage the majority of the land. Once they are gone, we will have lost an irreplaceable human resource.

Opportunities, Needed Explorations, and Outcomes

Given the changes that are taking place in the agricultural and food system, this is likely our last chance to develop effective strategies for regenerating a significant agriculture of the middle. The task before us is to frame a convincing rationale for a national initiative that will marshal the public and private resources to develop and test models as well as support existing ones for linking smaller, sustainable food enterprises on a regional basis and/or piggybacking such value chains on to existing distribution systems. These new food system approaches would explore and evaluate linkages between farms of the middle and corresponding enterprises of the middle in the rest of the food system—for example, regionally based food processors, distributors, and retailers. The new task will be to develop value chains that create a partnership among farmers, processors, distributors, and retailers based on a set of values that are tied to the products that the value chains produce.[20] Regionalism must be taken seriously in these explorations as the opportunities and constraints will differ among various regions of the country. And according to Elbert van Donkersgoed, value chains at their best consist of unique relationships in which "real partnership between all players in the chain. exists and relationships are built on trust."[21]   *regionalism*

Fortunately, we also have unprecedented opportunities to develop a food and farming system that can enable the agriculture of the middle to thrive. Already many small farms have demonstrated success with wood products, composting, agri-tourism, flowers, herbs, horticultural crops, nursery products, or wine. Midsize operations will be able to provide greater quantities of some of these products, and both small and midsize farms can be linked into marketing networks that can efficiently supply substantial quantities of these unique products.

Furthermore, a new market climate is emerging that will change the way we produce what we eat. This new climate, especially where food is concerned, consists of three distinct elements. Rick Schnieders, president and CEO of SYSCO Corporation, describes them as "memory, romance, and trust."[22] These are the attributes that an increasing number of food-conscious consumers are seeking. They want high-quality food, produced with farming practices that they want to support, and brought to them through a value chain they can trust. All of these attributes can be supplied readily and in sufficient quantity by the farmers and entrepreneurs who occupy the middle.

In fact, it may be that these unique food products can only be supplied in sufficient quantity by the farmers in the middle. The food service industry, which distributes the majority of these unique products, depends on the farmers of the middle to supply it. Again, as Schnieders points out, the needs of the food service industry are different than those of the traditional retail food market. Most food products in a typical retail grocery supermarket are manufactured from a few ingredients—notably corn, soybeans, sugar, and salt. The food service industry, on the other hand, supplies food operations that demand in excess of a hundred varieties of flour, not to mention thousands of food products with unique attributes, such as Vermont lamb, "antibiotic-free" meat, and Niman Ranch pork. The commodity market simply is not structured to provide such variety—its profits are based on the mass production of huge quantities of uniform commodities on narrow margins.[23]

importance of Food Service industry

Moreover, the climate of markets has recently changed. The concept of markets as conversations is an especially important one, imaginatively described in a book written by four authors, two of whom are on the management team of Sun Microsystems. The book is titled *The Cluetrain Manifesto*.[24] Markets, these authors contend, are undergoing a major shift as we enter the twenty-first century. During most of the twentieth century, markets consisted of "broadcast" information. If one wished to put a product on the market, one published a Sears, Roebuck and Company catalog, bought advertising in newspapers or magazines, purchased a spot on the radio, or bought time for an advertisement on prime-time television. Marketing involved one-way communication.

The authors of *Cluetrain* argue that the broadcast era is over, because twenty-first-century customers have grown up using the Internet and therefore are no longer receptive to having information broadcast to them. Today's customers are used to having a conversation about everything—including the products they buy and the food they eat. As such, anyone who does not provide an opportunity for customers to have a conversation about what they are selling will be at a distinct disadvantage in the marketplace. As the authors remind us, customers are not "seats or eyeballs or end users or consumers"; they are human beings whose reach exceeds our grasp.

What this analysis of the market of the twenty-first century tells us is that people increasingly will want to have relationships as part of their purchasing experience. Consequently, food marketers of the future who do not provide an opportunity for food customers to experience the story behind the food they buy are not likely to be in that market for long. This is the special magic behind today's direct-market success. When food customers go to the farmers market or buy from their local Community Supported Agriculture (CSA), they are buying a relationship as much as a food product.[25] The late Ken Taylor, founder of the Minnesota Food Association, used to describe this sort of relationship marketing in graphic terms. "People who live in urban communities for the most part don't like to get their hands dirty, but they surely want to shake the hand of someone who does."[26]

What are the implications of this transformation in the marketplace for the future of the agriculture of the middle? In the first place, this new development clearly gives the comparative advantage to precisely those farmers who are most threatened in the emerging two-part food system. Imagine a large number of small and midsize family farmers, linked together in a marketing network, producing food products for regional food sheds, using sound conservation practices, providing their animals with the opportunity to perform all their natural functions, preserving the identity of such food products by processing them in locally owned facilities, and making them available in the marketplace with opportunities for consumers to access the entire story of the product's life cycle using existing food service delivery systems.

Or imagine that a food service provider has a Web site listing thousands of food items with unique attributes and qualities. The chef in a restaurant could click on Vermont lamb, Niman Ranch pork, or Organic Valley cheese, and the Web site would link to a distribution network that would immediately place the order with the farmer or network of farmers who produce those unique products. Ten days later the product would appear at the chef's restaurant, together with the unique story of that product ready for the menu; the chef's credit card would be charged, and the farmer's account would be credited.

Models already are being developed by farmers and other food systems entrepreneurs that can provide a foundation for the national initiative we are proposing here. These models involve new enterprise structures and midtier value chains that can simultaneously serve the environment, rural communities, farmers, and the growing segment of the consuming public that wants to purchase foods with unique attributes. Midtier value chains are strategic alliances between independent (often cooperative) food production, processing, and distribution/retailing enterprises that seek to create and retain more value at the front end, and often operate on a regional level. Examples of these new midtier value chains include the wine industry in upstate New York, emerging alternative pork production and marketing systems in Iowa, and developing regional agricultural marketing labels (e.g., the Organic Valley Family of Farms in Wisconsin, PlacerGROWN in California, and Puget Sound Fresh in Washington).

It is this nation's larger small and midsize farms that have the comparative advantage in developing this new agriculture, since they have the flexibility to implement innovative production and marketing systems, and can produce the volume necessary to supply significant quantities of food into these new food chains.

The good news in all this is that we don't have to develop the markets for these new value chains. They are already there. Again, Schnieders asserts that the demand for differentiated products in the food service industry is large and the demand for sustainably produced foods is growing. Among other food customers, the health care industry has recently shown strong interest in acquiring more health-promoting foods for patients in hospitals and other care facilities. What is missing is a functioning value chain to get those products from the farmers to the consumers.[27] An extension system is also needed to assist farmers in making the transition from producing commodities to producing unique, differentiated products.

Farmers and other food entrepreneurs will not be able to develop such value chains by themselves. They lack sufficient capital and business experience. Furthermore, farmers have little experience in producing differentiated value products. They are experienced producers of undifferentiated commodities. It also would be helpful if there were changes in transportation and trade policies as well as meaningful constraints on the trends toward consolidation in the food and agriculture sector to reinvigorate significant free market competition in the food industry.

The research and education community must provide responsive leadership relative to analysis, model building, evaluation, and education in three areas: new production systems that meet the requirements of the emerging markets for highly differentiated products; market structures and relationships that link farmers producing these products with other food system entrepreneurs in a marketing value chain that enables farmers to produce and retain more value on the farm; and procedures and policies for recognizing, evaluating, and rewarding nonmarket benefits from the new agriculture as well as identifying and modifying policy structures that currently put smaller food enterprises and farmers of the middle at a competitive disadvantage.

It is the goal of the Agriculture of the Middle project to link the essential partners necessary to achieve these objectives. Our purpose is not to challenge or change commodity agriculture. Consolidation in the food and farming system is driven by powerful forces that are likely to play themselves out, and are probably largely beyond our control, especially as long as we continue to subsidize that system with public funds and ignore enforcement of antitrust laws. But as Michael Porter observes, there are two ways to be competitive in a global economy: by being the lowest-cost supplier of an undifferentiated commodity; or by providing the market with a unique and superior value in terms of product quality, special features, or after-sales service.[28]

Since only 10 percent of today's farmers produce more than 60 percent of the bulk commodities for the commodity market, it falls to 90 percent of the farmers, including those in the middle, to supply the other market. And it is precisely these farmers who are in the best position to produce the unique products demanded by that market. As numerous market analyses have shown, approximately 25 percent of present-day food customers want the unique products that this second market can offer them. One of the attributes that these markets increasingly want in their food choices is "locally grown by a family farmer."[29] And that market appears to be growing. So the markets and the producers are there. What is needed is the value chain to connect them.

A national task force consisting of farmers, university researchers, and food industry specialists first met in fall 2003 to begin addressing these issues. We are: identifying how best to structure and fund the research that can start to identify the opportunities and barriers involved in developing such value chains; determining how to provide educational opportunities for farmers, officials, and the public as to the importance of the agriculture of the middle along with what should be done to preserve it; and developing the structure to make this marketing middle a reality. We will devise a plan to obtain a substantial commitment to public-sector research and education, and private-sector partnerships that can evolve food systems approaches to revitalizing the agriculture of the middle through such market opportunities.

The immediate, specific outcome we expect from this effort will be a plan of action to secure the support to carry out this ambitious agenda. If we are successful in that effort, the long-term outcomes we envision include the following:

• The development of more comprehensive, regionally appropriate, ecologically sound agricultural production systems that enable farmers of the middle to produce and retain more value on their farms while

restoring the health of local ecosystems and contributing to the revitalization of rural communities.

• The creation of new market structures/models and marketing relationships for midsize farms that create and retain greater value in the farm
and rural community sectors, and that increase the viability of local and
regionally based food processing and distribution enterprises, and/or develop such value chains within existing enterprises.

• The exploration of policy alternatives that support these new marketing and production systems.

• The education of a large number of consumers in the market of the
middle who are aware of the contributions of the agriculture of the middle, and support these farmers with their food choices and purchases.

• The development of a national cadre of researchers and food system
practitioners with expertise and commitment to microenterprise food
system analysis and reform.

• The assurance that the information and other inputs needed for
healthy, diverse food systems remain in the public domain.

It is critical to remember that none of this can happen apart from sustaining a particular kind of farmer with a particular kind of farm. Of all
the millions of words that have been written about agriculture over the
last fifty years, perhaps none have described what we need more elo-
quently than those of Wendell Berry more than a decade ago:

If agriculture is to remain productive, it must preserve the land, and the fertility
and ecological health of the land; the land, that is, must be used well. A further
requirement, therefore, is that if the land is to be used well, the people who use it
must know it well, must have time to use it well, and must be able to afford to
use it well. Nothing that has happened in the agricultural revolution of the last
fifty years has disproved or invalidated these requirements, though everything
that has happened has ignored or defied them.[30]

Notes

Parts of this chapter are reprinted from Fred Kirschenmann, "The Current State
of Agriculture: Does It Have a Future?" in *The Essential Agrarian Reader*, ed.
Norman Wirzba (Lexington: University of Kentucky Press, 2003). A draft of
this chapter was placed on the ⟨http://www.agofthemiddle.org⟩ Web site to in-
vite interested parties to become involved. Numerous people provided additional
information, proposed deletions and alternatives, and suggested rewrites. We
have incorporated many of these helpful suggestions into this chapter.

1. See Willard W. Cochrane, "A Food and Agricultural Policy for the 21st Century" (unpublished paper available from the author, 1999). Cochrane points out

that as of 1997, over 61 percent of the total agricultural production in the United States came from just 163,000 farms, and that 63 percent of those farms were already producing a single commodity under contract with a consolidated firm. Meanwhile, 575,000 small- to medium-size family farms generate 30 percent of the total national production.

2. In this chapter, we are using the term midsize farms to describe those farms that the U.S. Department of Agriculture's (USDA) collapsed farm typology calls "intermediate farms"—farms with gross annual sales of between $100,000 and $250,000, and where farming is the primary occupation of the owner(s).

3. Economic Research Service, "Farm Typology for a Diverse Agriculture," USDA-ERS Agriculture Information Bulletin no. 759 (Washington, DC: ERS, September 2000).

4. Patrick Martins, "Set That Apricot Free," *New York Times*, April 24, 2004.

5. Calvin Beale, "Salient Features of the Demography of American Agriculture," in *The Demography of Rural Life*, ed. David Brown et al., publication no. 64. (University Park, PA: Northeast Regional Center for Rural Development, 1993).

6. Reported by Fred Kirschenmann, who attended the meeting.

7. USDA, *A Time to Choose: A Summary Report on the Structure of Agriculture* (Washington, DC: USDA, January 1981), 147, 150.

8. See the work of William Heffernan et al. tracking the consolidation in the agriculture and food industry for the past twenty-five years. See, for example, "Consolidation in Food Retailing and Dairy: Implications for Farmers and Consumers in a Global Food System" (January 8, 2001), available from the National Farmers Union.

9. W. C. Lowdermilk, "Conquest of the Land through Seven Thousand Years," USDA, Soil Conservation Service, Agriculture Information Bulletin no. 99 (1953).

10. Michael Boehlje, "Structural Changes in the Agricultural Industries: How Do We Measure, Analyze, and Understand Them?" *American Journal of Agricultural Economics* (December 1999): 1028–1041.

11. "Arkansas Pecking Order," *Time*, October 26, 1992, 54.

12. David W. Orr, "The Events of 9-11: A View from the Margin," *Conservation Biology* 16, no. 2 (April 2002): 289.

13. See, for example, David Tilman, "The Greening of the Green Revolution," *Nature* 396 (November 19, 1998), and "Biodiversity and Ecosystem Functioning: Current Knowledge and Future Challenges," *Science* 294 (October 26, 2001): 804–808.

14. Quoted in William Greider, "The Last Farm Crisis," *Nation*, November 20, 2000, 16.

15. Edward O. Wilson, *The Future of Life* (New York: Alfred A. Knopf, 2002).

16. Ibid., 130.

17. Wendell Berry, *Another Turn of the Crank* (Washington, DC: Counterpoint Press, 1995), 11.

18. Lewis Mumford, "Authoritarian and Democratic Technics," in *Questioning Technology*, ed. John Zerzan and Alice Carnes (Philadelphia, PA: New Society Publishers, 1991).

19. R. Edward Grumbine, *Ghost Bears* (Washington, DC: Island Press, 1992).

20. A value chain is a network of collaborating players who work together to satisfy market demand for a specific product and/or set of services.

21. Elbert van Donkersgoed, "Value Chains versus Supply Chains," *Corner Post*, June 23, 2003, 2.

22. Rick Schnieders, keynote address (delivered at the Practical Farmers of Iowa annual meeting, January 25, 2003).

23. Rick Schnieders, "The Strategic Role of the General Counsel at SYSCO" (speech delivered to the General Counsel Institute at Georgetown University, April 2004; available from the author).

24. Christopher Lacke et al., *The Cluetrain Manifesto* (Boulder, CO: Perseus Books Group, 2000).

25. CSA refers to a community of individuals who pledge support to a farm operation so that it becomes the community's farm, with the growers and consumers providing mutual support as well as sharing the risks and benefits of food production. Typically, members or "shareholders" of the farm pledge in advance to cover the anticipated costs of the farm operation and farmer's salary. In return, they receive shares in the farm's bounty throughout the growing season. Members also share in the perils of farming, including poor harvests due to unfavorable weather or pests. This provides the farmers with working capital in advance, thereby relieving them of much of the burden of marketing.

26. Personal conversation.

27. Schnieders, "Strategic Role of the General Counsel."

28. Michael E. Porter, *The Competitive Advantage of Nations* (New York: Free Press, 1990), 37.

29. Rich Pirog, "Ecolabel Value Assessment Phase II: Consumer Perceptions of Local Foods," Leopold Center for Sustainable Agriculture, 2004, available at ⟨http://www.leopold.iastate.edu⟩.

30. Wendell Berry, *What Are People For?* (San Francisco: North Point Press, 1990), 206–207.

2

Pursuing Long-Term Food and Agricultural Security in the United States: Decentralization, Diversification, and Reduction of Resource Intensity

Kenneth A. Dahlberg

[handwritten: What are the risks of centralization]

Any attempt to establish and sustain an agriculture of the middle will benefit from seeking to understand the risks that today's societies face due to their increasing centralization, simplification, and resource intensity. Centralization is a widely used, but not fully understood concept. Historically, it has resulted from the accretion of infrastructures that are progressively larger in scale and scope (roads, aqueducts, irrigation systems, railroads, telecommunications, etc.). Their interactions with existing institutions have led to the creation of new, more centralized organizations. Over human history, this is seen in the significant changes associated with the "grand transitions" from hunter-gatherer societies, to agricultural societies, to city-based civilizations, to modern industrial society.

[handwritten right margin: bigger infra-structures where our course will begin w/]

The simplification of societies that has accompanied this process is much less understood—particularly because of a common conceptual confusion between systems that are "complicated" and those that are "complex." Ecosystem theory defines the complexity of a natural system in terms of the diversity of its distinct species (those that cannot reproduce one with the other), and the extent and patterns of their spatial distribution. There is an ongoing debate in ecosystems theory as to whether or not complex ecosystems have greater resiliency than simple ones. Resiliency refers to the ability of a system to rebound from major disruptions that exceed those found in its normal operating environment. In contrast, it clear that *managed* ecosystems—which have increased exponentially over time—become less resilient as they are simplified and made more resource intensive (Vandermeer et al. 2002).

Modern societies have been simplified as industrial systems have increased their resource intensity and have increasingly applied mass production techniques, particularly functional specialization and standardization. As these techniques are applied to more sectors and levels of society, complexity is reduced, as is resiliency—which is the price of the greater productivity achieved. Thus, industrial societies are more productive and more complicated than earlier ones, but are less complex and less resilient to large-scale disruptions—whether natural, social, economic, or political.

The risks of such disruptions become greater as more and more humans (and livestock) increase their appropriation, exploitation, and degradation of nature through their growing consumption of ever more resource-intensive products and foods. Losses of cultural diversity and biodiversity as well as increasing greenhouse gases, global warming, and freshwater contamination threaten not only modern societies but the very sources of life.

The Simplification and Centralization of U.S. Society

Those who settled the United States brought with them the many traditional as well as modern beliefs that helped shape the infrastructures and institutions of this "new" land. These included Judeo-Christian beliefs in the separation of humans from nature. Two other eighteenth-century beliefs—that functional specialization is the best technique for organizing production, and that technologies are inherently neutral—provided important rationales for Alexander Hamilton's economic vision of a manufacturing-based society as compared to Thomas Jefferson's democratic vision of the United States as a diverse combination of small-scale and largely self-reliant farmers and artisans.

Much more significant, however, was the nineteenth-century expansion in the United States of corporate powers through the court-created legal fiction that corporations are "persons" with the same constitutional and legal rights as individuals (Grossman and Adams 2001). The increasing growth of centralized bureaucracies along with the standardization of corporate, government, and nonprofit job categories has meant that the initial organizational basis of modern industry—functional specialization—has been spread across all sectors of society.

This increased societal centralization has generated three mutually reinforcing risks: the loss of democracy, the loss of cultural diversity and biodiversity, and the loss of our society's capacity to adapt to major changes and/or disruptions. The losses of democracy to the unchecked power of complexes such as the military-industrial complex, the oil-auto-highway complex, and the agrichemical-agribusiness-grain export complex also reduce society's adaptive capacities. Since industrial societies are structurally more subject to disruption, they need more, not less adaptive capacity. In systems terms, traditional checks and balances as well as other democratic structures can be seen as the negative feedback loops required to enable government—one of the central institutions of society—to have an ongoing adaptive capacity to deal with change.

The losses of cultural diversity and biodiversity caused by modern centralized systems in turn seriously threaten them. This is because the single-minded promotion of industrial production and trade results in costly pollution, toxic wastes, loss of habitat, and increased global warming and climate change. More fundamentally, the global loss of cultural diversity and biodiversity is reducing the capacity of social and natural systems to adapt to, and buffer, the increasing pressures of industrialism and global warming as well as to continue to provide invaluable ecosystem services. Since the loss of adaptive capacity has been the cause of the collapse of almost all previous societies, we must preserve and enhance those natural and social resources nourishing that capacity.

It is only by transforming the *institutional sources* of the centralization of power that we will have a chance of doing this, and adapting and muddling through to a post–fossil fuel era. Thus, while we need to take measures to protect ourselves and current institutions, we must do that within the context of a much larger and ambitious strategy to restructure and make less resource intensive our overcentralized infrastructures, reform what are now maladaptive institutions and overconcentrated economic sectors, and strengthen our adaptive capacities as a society. It needs to be stressed here that the goal is not a total decentralization but a rebalancing that removes excessive power from higher levels, while leaving (or adding) necessary (but not unchecked) powers at other levels of government and society. If we do otherwise and focus only on trying to defend current—and unsustainable—infrastructures and institutions, we will lose our adaptive capacity entirely, and commit ourselves to a chaotic and destructive future.

The accelerating and mutually reinforcing centralizing trends since World War II have fostered increasing concentration in U.S. agriculture, where oligopolies exist in virtually every input and output sector (Heffernan et al. 1999; ETC Group 2003). The focus of all major stakeholders in agriculture on production efficiency and specialization rather than systems approaches has been a major source of this concentration. The lack of effective antitrust enforcement and the push for the export of agricultural commodities—led by large agribusiness firms like Cargill, ADM, and others—have also significantly contributed to a corporately controlled and managed food system.

The Vulnerability and Risks of the Three Main Types of U.S. Agriculture

The three main types of agriculture in the United States today— agribusiness, agriculture of the middle, and alternative agriculture—are different in terms of both their vulnerability and the risks they pose to society. The long-distance, high-tech agribusiness food system is the most homogeneous of all the systems, and the most vulnerable to any major disruption of its natural and/or social resources. Significant increases in energy prices will dramatically affect it—since they will have a multiplier effect throughout its entire input, output, processing, distribution, marketing, and waste streams. It is also the system most vulnerable to a range of economic and social disruptions—from the loss of commodity and export subsidies (likely from international trade pressures combined with extreme domestic budget pressures); the loss of international markets as more countries demand the labeling of genetically modified organisms (GMOs) and/or more organic production; increased pressures from large processors and retailers to cut costs; possible future reformist presidents and/or congresses; and wider public pressure to make agriculture more genuinely sustainable and place it in the larger framework of public health.

Beyond such vulnerabilities to economic, social, and political changes and disruptions, long-distance, high-tech agribusiness is even more vulnerable to disruption and collapse triggered by major calamities such as climate changes, major accidents, and/or terrorist attacks. Structures and systems that are inherently highly vulnerable due to their structure cannot be effectively protected in the longer term. Just as our larger society, agribusiness faces the challenging choice of either addressing its real long-term vulnerabilities by making genuine reforms or continuing on a path leading toward collapse.   *greater adoptive capacity of the middle*

The agriculture of the middle food system is less homogeneous in structure than long-distance, high-tech agribusiness, so there are a wider range of potential vulnerabilities, impacts, and risks. The choices involved in addressing them—still difficult—will be somewhat more varied; they also will be helped by somewhat greater adaptive capabilities. Agriculture of the middle farmers can continue to align themselves with agribusiness, or seek out new and distinct paths—paths that seek different, less resource-intensive, more systems-based approaches according to region, type and size of farm, and so on. To the degree that agriculture of the middle continues to follow long-distance, high-tech agribusiness, it will face both the vulnerabilities of agribusiness as well as a continuing decline in the number of farms, farm families, and rural communities. This choice will also threaten the health of rural America, its landscapes, and biodiversity more generally. The other choice—that of seeking to combine what was best from its past with the needs and opportunities of pursuing a more regional and sustainability-based approach—is challenging, but more adaptive.

The alternative agriculture food system—so labeled for lack of a better term—is the most heterogeneous of the three modern industrial systems. It includes a variety of systems-based approaches (organic, agroecological, biodynamic, and sustainable food systems) as well as a variety of foci (community food security, local food policy councils, local/community development, social justice, public health, etc.). Many involved in the alternative food system are small-scale and local farmers seeking to build more localized economies through CSAs, farm-to-school programs, local processing, and so forth. The research to date shows that alternative agriculture takes many different forms in different regions and locales.

Given a stronger and more diverse local base, alternative agriculture may not be as immediately vulnerable to the types of disruption outlined above. Yet the ever-increasing pressures for structural simplification and standardization flowing from corporate and governmental centralization—plus the preemption by trade treaties of national, state, and local powers to regulate their respective health, safety, and environmental standards—will make it extremely difficult in the long term for localities to create and maintain genuine alternatives to industrial food and agriculture. Thus, the healthy growth or decline of alternative agriculture can be seen as a bellwether for the larger risks that industrial food and agriculture systems pose for U.S. society at large.

Emerging Risks

Over the next decade, three emerging risks are visible. First, there is the risk that agriculture will be redefined in terms of national security, thereby losing its soul. The leading edge of this redefinition is the push to extract biofuels from cropland—all in the name of energy independence. To the degree that agriculture becomes a fuel supplier for the national (read global) market, it will be opening the door to its also becoming a major supplier of industrial feedstocks for plastics, pharmaceuticals, and other industrial products. Farmers will become "miners of the soil," and will have as little power or independence as coal or other hard rock miners.

The public health risks of overcentralized food and agricultural systems are more visible, and will become increasingly important. The most significant public health risk is likely to be the emerging crisis of obesity that has resulted from a combination of cheap foods generated by commodity subsidies and the "supersizing" of food products—something driven by the need of the food industry to continually show quarterly earnings growth. This is a theme that Marion Nestle (2003) has stressed. Since only 20 percent of the industry's costs are at the farm gate, supersizing is the easiest way to increase profits since it costs relatively little to increase the size and price of food portions. Also significant are the increasing health costs of the growing number of antibiotic-resistant strains of disease that are caused by highly centralized confined livestock feeding operations using huge amounts of antibiotics to control animal diseases (Khachatourians 1998; WHO 2002).

Third, to the degree that centralization and resource intensification continue, the regenerative base of agriculture will be increasingly threatened as exemplified by increasing losses of livestock biodiversity, with poultry being a prime example (Pisenti et al. 1999). More broadly, crop varieties will continue to vanish, precisely at a time when the pressures of global warming, increasing weather extremes, and increasing crop and animal diseases—exacerbated by monocultural and confinement production practices—call out for greater diversity, decentralization, and regenerative capacities.

Pursuing Genuine Food Security in the United States

dam analogy ✓

The discussion so far has stressed the growing vulnerability of industrial societies due to their ever-increasing centralization, and the risks and costs flowing therefrom—especially the loss of democracy, cultural and biological diversity, and adaptive capacity. The current industrial food and agriculture system can be conceptualized as a gigantic irrigation system, where a small number of people (producers and suppliers) control all the large dams and their flows (pools of resources) as well as most of the downriver operators who allocate regional and local flows. New adaptive strategies for the agriculture of the middle—namely, decentralization, diversification, and democratization—involve changing the management at each level (democratization), removing oversize dams and replacing them with more small and midsize ones that better fit the topography and climatic zones (decentralization), plus building new canals, networks of ditches, and water conservation systems to ensure the availability of water to smaller growers and towns (diversification).

The overall strategy is to change control over food flows; change the built infrastructure so that the flows are more widely distributed; and design new infrastructures and conservation measures that fit into and help reestablish the natural flows of the underlying ecological and social systems. By their nature, infrastructural and institutional reforms to the food system are long-term and difficult objectives. Subsidies to farmers should ideally be scale- and commodity-neutral ones. Leveling the playing field in food and agriculture would dramatically encourage a regionalization and localization of agriculture and food systems.

Pursuing infrastructural reform will require major corporate reforms, however, particularly in terms of reducing the high levels of oligopoly in both agriculture and food (Heffernan et al. 1999; ETC Group 2003). Shorter-term measures to complement antitrust and corporate reform include: making sure livestock markets are genuinely competitive through bans on packer ownership of livestock; legal reforms that make livestock corporations—not contract farmers—responsible for health and environmental problems; rebuilding local livestock infrastructures (milk processors, abattoirs, distribution centers, etc.); and removing exploitative tax shelters (Benbrook 2003).

Diversifying food systems will require efforts on a number of fronts: strengthening enforcement of existing air, water, wetlands, forest, and endangered species legislation; getting rid of subsidies that favor particular regions with cheap water and/or transportation; and developing new policies such as a carbon tax to level the energy playing field. Systems-based regulation will require an underpinning of research that examines long-term impacts in various systems and sectors. Here, the fragmentation of governmental and university research along agency and disciplinary lines remains a formidable obstacle.

Among the more innovative proposals relating directly to food and agricultural diversification are those of Charles Benbrook (2003). His proposals include: basing federal farm program payments on per acre nitrogen uptake efficiency, coupled with the diversity of rotations; and spreading livestock out across the cultivated cropland base by reducing western water subsidies and ensuring that there is an accessible USDA-certified abattoir in every agricultural county. These and other measures would relocate middle-level production systems regionally to better fit their underlying habitats and environments by moving most of the beef and dairy industries to the upper Midwest and New England, and poultry and hogs west from the Piedmont region—thus also reducing resource intensity greatly. New approaches to farmland preservation are also needed—approaches that would link the purchase of development rights from farmers to their retirement benefits. State laws that restrict corporate ownership of farms, such as those in the Dakotas and Nebraska, deserve serious consideration in other states. There are a host of other federal and state laws that need to be reviewed in terms of their impact on an agriculture of the middle.

Genuine food and agricultural security will result from decentralized and diverse systems because their lower resource intensity makes them inherently more stable and resilient. As they seek to revitalize an agriculture of the middle, farmers and rural communities must view current U.S. approaches to homeland security with caution. Just as U.S.

approaches to national security need to be redefined as well as moved on to firmer long-term conceptual and organizational grounds, so too must homeland security approaches to food and agricultural systems. Table 2.1 illustrates the basic, longer-term changes that are needed to make U.S. agriculture more secure.

Central to the needed approaches and long-term security is the development of a new and food system specific understanding of sustainability. One recent attempt is the following:

Sustainability as it applies to food means that societies pass on to future generations all the elements required to provide healthy food on a regular basis: healthy and diverse environments (soil, water, air, and habitats); healthy, diverse, and freely reproducing seeds, crops, and livestock; and the values, creativity, knowledge, skills, and local institutions that enable societies to adapt effectively to environmental and social changes. (Dahlberg et al. 2005)

Strategic Implications

how will the 3 types respond?

Successful efforts to create true long-term food security by decentralizing, diversifying, and placing U.S. food and agricultural systems within a genuine homeland security framework will depend in large part on how the three different types of agriculture respond to larger societal challenges as well as each other. It will also depend on how a range of groups concerned about food and agriculture frame and pursue reforms.

In light of the above analysis, agribusiness can be seen as in a position similar to the large trusts of the nineteenth century, and can be expected to strongly resist efforts to reduce its powers, subsidies, and privileges. Thus, one of the most important tasks overall will be educating and persuading farmers of the middle that the sooner they make a commitment to more sustainable, diverse, localized, and less resource-intensive approaches to agriculture, the easier—and less risky—their personal transition will be. *farmers of the middle were sold on these ideas*

Since World War II, farmers of the middle have been sold the idea that pursuing modern industrial, high-input, long-distance approaches was in their interest. Of course, those selling such ideas—large input suppliers, large banks, large processors, exporters, and sadly the land grant system—had even more to gain than the farmers. Just like the many small businesspeople who support large national lobbying organizations that pursue policies that benefit big businesses at the expense of small

Table 2.1
Contrasting Homeland Security Approaches to Agriculture

	Current approaches	Needed approaches
Infrastructure	Overcentralized; functionally specialized; complicated; resource intensive	Decentralized; systems-based; complex and diverse; less resource intensive
Policy approach	Top-down; standardized; control; secrecy; technical fixes	Democratic at each level; diverse; transparent; socio-regulatory
Environment	Separate rules and regulations for air, water, land—but based on public health or species protection	Integrated rules and regulations for specific habitats, watersheds, food sheds, etc., with stronger public health and biodiversity protection
Health	Medical response model (individual and disease focus) for both humans and animals	Public health model (preventive; focus on promoting healthy populations and environments)
Agriculture	Focus on individual farms; pesticides, herbicides, pharmaceuticals, and quarantines	Focus on healthy farms, families, towns, and regions; sustainable production systems through a diversity of crops, species, and landscapes
Food safety	Specialized by product; inspections and recalls (mainly postproduction)	Systems and process safety; inspections and recalls (mainly before and during production)
Biofuels	Mine the soil for energy; redefines agriculture within a national security framework; will reduce cropland for food and increase food prices	Sustainable biofuels for local needs; part of building local and regional self-reliance; does not compete with cropland for food
Industrial feedstocks	Convert crops to plastics, pharmaceuticals, and other industrial products; will further place farmers (miners of the soil) at risk	Reinvigorate Jeffersonian understandings of farming as a way of life, a bulwark of democracy, and as providing true homeland security

ones, small farmers have supported farm organizations like the Farm Bureau that mainly benefit large, resource-intensive farmers. Farmers of the middle will need to be educated and persuaded that their future no longer lies in seeking to keep up with the agribusiness "leaders" on the industrial production treadmill; rather, they need to reexamine and rethink their approach to agriculture. In this they have a great deal to learn from alternative agriculture. *persuade/educate farmers*

The most critical challenge that farmers of the middle face is to avoid the temptations of joining high-tech agribusiness in becoming the suppliers of biologically based feedstocks for the energy, plastics, and pharmaceutical oligopolies that are now seeking cheaper replacements for fossil-fuel-based feedstocks. The economic, political, and environmental costs of small and midsize farmers becoming miners of the soil need to be made clear. On the most fundamental level, farmers of the middle need to see that they risk losing their individual and collective agricultural souls if they follow this path. Choosing the other path will be difficult, but it will be their salvation in the longer term. It will require a revitalization of those traditional values that are adaptive as well as a conscious commitment to search for and define what it truly means and takes to be a family farmer and citizen in this time of transition. Individually and together, family farm members need to lead in finding new and sustainable ways to nourish the soils on their farm; enhance its waters, air, and habitats; create and maintain healthy working and home environments; contribute to their local community and region; and raise healthy food for people. As citizens, they will need to help society develop the new values, appropriate technologies, and more decentralized and democratic institutions that will be required as we move into the post–fossil fuel era.

References

Benbrook, Charles. 2003. What Will It Take to Change the American Food System? Paper presented at the Kellogg Foundation Food and Society Networking conference, Woodlands Center. Available at ⟨http://organic.insightd.net/reportfiles/Kellogg_ChangetheSystem_April_2003.pdf/⟩.

Dahlberg, Kenneth, Jim Bingen, and Kami Pothukuchi. 2005. The Albion Statement. Healthy People, Places, and Communities: A 2025 Vision Statement for

Michigan's Food and Farming. Available at ⟨http://www.mifooddemocracy.org/⟩.

ETC Group. 2003. Oligopoly, Inc. Concentration in Corporate Power: 2003. *Communiqué* 82. Available at ⟨http://www.etcgroup.org/⟩.

Grossman, Richard L., and Frank T. Adams. 2001. Taking Care of Business: Citizenship and the Charter of Incorporation. In *Defying Corporations, Defining Democracy*, ed. Dean Ritz, 59–72. New York: Apex Press.

Heffernan, William, Mark Hendrickson, and Robert Gronski. 1999. Consolidation in the Food and Agriculture System: Report to the National Farmers Union. Available at ⟨http://www.foodcircles.missouri.edu/⟩.

Khachatourians, G. G. 1998. Agricultural Use of Antibiotics and the Evolution and Transfer of Antibiotic-Resistant Bacteria. *Canadian Medical Association Journal* 159:1129–1136.

Nestle, Marion. 2003. The Ironic Politics of Obesity [editorial]. *Science* 299:781.

Pisenti, Jacqueline M., et al. 1999. Avian Genetic Resources at Risk: An Assessment and Proposal for Conservation of Genetic Stocks in the USA and Canada. Report no. 20. Davis: University of California Davis Division of Agriculture and Natural Resources, Genetic Resources Conservation Program.

Vandermeer, John, Deborah Lawrence, Amy Symstad, and Sarah E. Hobbie. 2002. Effect of Biodiversity on Ecosystem Functioning in Managed Ecosystems. In *Biodiversity and Ecosystem Functioning: Synthesis and Perspectives*, ed. Michel Loreau, Shahid Naeem, and Pablo Inchausti, 221–233. New York: Oxford University Press.

World Health Organization (WHO). 2002. Use of Antimicrobials outside Human Medicine and Resultant Antimicrobial Resistence in Humans. Fact sheet no 268. Available at ⟨http://www.who.int/mediacentre/factsheets/fs268/en⟩.

II

Organizational Structures That Could
Support an Agriculture of the Middle

3

Cooperative Structure for the Middle: Mobilizing for Power and Identity

Thomas W. Gray and G. W. Stevenson

cooperatives

This chapter is about cooperatives and the application of their structures to the national agriculture of the middle initiative. Cooperatives can be considered formal collective actions related to social movements. Collective actions and social movements come out of preexisting socioeconomic and historical contexts (and larger societal dissatisfactions and dissent). Cooperatives are unique among economic organizations in that they are at once democratic associations of members as well as businesses. This dual character of cooperatives generates potentially creative tensions within the organization that are shaped as the cooperative interacts with its environment.

As defined in the literature, "cooperatives are user owned and controlled businesses from which benefits are derived and distributed on the basis of use" (Dunn 1988, 85). This is distinct from investment-oriented firms, wherein investors with money seek to make a return on that money by investing in an activity that will return a profit. Members (or potential members) of a cooperative organize collectively to provide some service (broadly defined) to themselves. This can be a range of services from marketing food products, to purchasing agricultural input supplies, to the bulk purchasing of groceries. A financial return above costs (termed margins in cooperative language) must be made to continue to provide the service.

The agriculture of the middle initiative is a multicentered ensemble of interests loosely organized around the sociological, economic, and ecological concerns of sustainability related to midsize farms, ranches, and fisheries, which have difficulty marketing food products directly to local customers or selling agricultural commodities through increasingly global marketing structures (see chapter 1; Lyson and Green 1999).

"Middle interests" have two fundamental organizing tasks: to build economic viability for midsize agrifood enterprises while holding their ground against those powerful socioeconomic dynamics—corporate conglomeration, constructed consumerism, and elite globalization—that create redundancy among farmers and communities, and to build such economic viability in a manner consistent with the values and goals of sociological and environmental sustainability.

Struggles Related to Power and Identity: "To Have" and "To Be"

Social movements and collective actions have evidenced historical distinctions between "having and power," and "being and identity" (Melucci 1988, 1996). Older labor and cooperative movements focused more on struggles over power and "getting a fair share," while many contemporary movements are oriented more to being and identity. The agriculture of the middle initiative, as a movement, tends to reflect both struggles, plus it calls for fuller democratization. The following section will highlight these considerations in light of historical trends and dynamics (see Fairbarin 1999, 2004; Lyson 2004; Gray et al. 2001; Sexton 1997).

Given the persistence and invasiveness of agribusiness power, agriculture of the middle farmers will likely continue to struggle for fairer distributions of resources, fairer prices, and greater power in the marketplace. Pressures on farmers to increase their scale and industrialize (or sell out) will likely continue. Multiregional and multinational firms can be expected to continue expanding and solidifying their market positions via integration, conglomeration, and the worldwide sourcing, processing, and selling of food products. Originating in demands for a short-term return on investment, much of this pressure will be felt in the marketplace through pricing strategies and trading relationships. Predictability and the control of risk will be maximized in transactions with farmers via power relationships, perhaps most solidified through production contracts. Production contracts are legally binding agreements between a farmer and buyer concerning various standards of production, product, and delivery—with most of the risk falling on the farmer (Thu and Durrenberger 1997).

These factors tend to push struggles over power, fairness, and surviv-ability to concrete levels. The logic of an exclusive instrumental rational-ity tends to transform the character and culture of family farming, from diverse and independent entrepreneurs, to cogs in a machine for making short-term return on investment for large agribusinesses. Struggles over both fairness and economic survival, and identity and meaning (what it means to be a family farmer as a nurturer of land, livestock, and commu-nity), become sharpened and more obvious.

To survive as producers, agriculture of the middle farmers will need market power to countervail these trends and tendencies. Historically, many agricultural cooperatives were organized to oppose monopoly and oligopoly investment firms on the local, regional and national levels. John Craig (1993) argues that co-ops have been instrumental in breaking monopolies and cartels, eliminating windfall profits, eliminating middle-people, and providing for more equitable distribution of wealth. Though not the dominant form of agribusiness in the early twenty-first century (except in a few commodities), the agrifood market share for coopera-tives is usually about one-third of marketed food products and over one-fourth of agricultural input supplies (USDA 2004). From a historical point of view, Patrick Mooney writes (2004, 78), this must be recognized as success, given the origins of the cooperative movement as a form of resistance to large power disparities between farm and firm, and given cooperative resilience in continuing to service farmers over time in the face of this power.

It needs to be noted, however, that these older cooperative associa-tions were formed in an era when mobilizations were organized predom-inantly for power and getting a fair share. Many are rooted in the first half of the twentieth century, when words like "ecology" and "sustain-ability" were barely in the language. As mentioned, collective mobiliza-tions and "new social movements" within the socioeconomic culture of high modernity tend more often to be grounded in concerns of identity, safety, a sense of permanence, and broader democratization of or oppo-sition to unaccountable power (Melucci 1988, 1996; Buechler 1995; Larana et al. 1994; Johnston et al. 1994). Farmers and ranchers are not separate from these larger socioeconomic cultural influences as they are filtered through the agrifood context.

Older cooperative movements held answers for farmers as a counter-vailing power in the marketplace. To be successful, agriculture of the middle interests must forge strategies that encompass not only market power and economic survival but also identities that highlight sustain-ability along with the permanent importance of healthy ecological and community environments. Agriculture of the middle is also a voice for greater democratization in that it represents an alternative to the power-ful dynamics of farm concentration, farmer displacement, and market dominance by increasingly large agribusiness powers. Its very goals in-volve democratization in that it seeks to create conditions for dispersed control of land, resources, and capital. And it opposes the predominant business logic—a return on investment—that tends to exclude larger so-cietal costs and interests. Agriculture of the middle identity touches on all these issues, and each has implications for mobilization.

Cooperative Organizations: Empowerment and Identity Vehicles

Cooperatives have much to recommend themselves as empowerment and identity vehicles for agriculture of the middle farmers and ranchers. They are democratic and business organizations organized not around a short-term return on investment but three general principles of use:

1. The user-owner principle: Those who own and finance the coopera-tive are those who use the cooperative.
2. The user-control principle: Those who control the cooperative are those who use the cooperative.
3. The user-benefits principle: The cooperative's purpose is to provide and distribute benefits to its users on the basis of their use. (Dunn 1988, 83)

For cooperative members, their participation in the activity of the co-op is central to their relationship with the organization. The members are the users as well as the owners and governors of the firm. (This is unlike investment firms, wherein investor-owners have little connection to the business activity, beyond a return on their investment. If investor-owners make use of the business activity, or even its products, it is only on an incidental basis.) The internalization of relationships (owners, users, and governors) within a cooperative organization creates the potential for members to develop an organization that is of their liking or at least within their influence. Power is not anonymous; the cooperative structure is transparent and creates potentials for significant accountability.

As an organizational democracy, cooperatives provide a forum for working out differences and conflicts, coming to negotiated solutions that can permit collective actions, while allowing for a continuing diversity among the membership. Several different agendas can be accommodated, or at least entertained and discussed. Mooney (2004) suggests the cooperative form may be ideal for accommodating the many interests of "sustainable" development since its structure is designed to bring together the sociological with the economic via a democratic process.

The democratic form of cooperatives may itself help solidify an agriculture of the middle collective identity. Participation in cooperative organizations can bring an awareness (even to reluctant members) that they can achieve collectively, what they cannot do individually. This may include ecological, economic, and community goals. This can in turn deepen an appreciation of their mutual and joint interests. The more members find their individual identities in the shared experience of the group, the greater the solidarity of the group. The greater the solidarity, the greater the commitments to collective goals and the willingness to subordinate individual interest to the overall collective action (Johnston et al. 1994; Melucci 1996).

Mutuality between Agriculture of the Middle and Consumer Interests

The overall viability of agriculture of the middle interests depends on their being responsive to emerging markets for food products. The evidence is mounting that consumers are increasingly demanding highly differentiated and values-related food products, especially in the food service industry. Restaurants, health care facilities, schools, other food service enterprises, and some supermarkets increasingly are demanding foods that:

• have superior taste, health, and nutritional qualities
• are associated with unique food stories that identify where the food comes from and how it is produced
• come to them through transparent supply chains built on business relationships they can trust and support (see chapter 7 on midtier food value chains)

These food products and relationships are much more than those embedded in the bread and butter, milk, meat and potatoes of an earlier era. Agriculture of the middle farmers and a growing number of consumers have much in common when understood as residing under the general influences of a high modernity culture. Needs for safety, a sense of permanence (sustainability), and demands for meaning, self-determination, and resistance to unaccountable power are increasingly shared. Middle farmers and ranchers will need to combine strategies for empowerment in the marketplace (to have) with strategies for deepening an agriculture of the middle identity (to be) that engages definitions of family-based agriculture as necessary for ecological, economic, and community well-being. Consumers will need to become "food citizens" (Miller 1995; Stevenson 1998), making choices in both the market and political sectors that are supportive of parallel economic, ecological, and community values. Such mutually supportive development by farmers and ranchers as well as consumers can make for a powerful political and cultural economy of food.

Cooperative Structure: Locals, Centralized Cooperatives, and Federations

We believe that the best linking device between farm field and grocery cart is cooperative organization. This observation, however, is made with provisos. Being dual in nature—both democracies and businesses—cooperatives have embedded within them values of equality, equity, participation, and self-governance, but also efficiency, performance, and economic return. Depending on its competitive environment, incremental changes can occur that shift a cooperative organization between managerial expertise (and demands for a return on investment) and membership needs, efficiency and equality, authoritarian logic and democratic logic, and bureaucracy and democratic participation. These are inherent tensions within a cooperative.

Care needs to be taken in a cooperative not to inordinately privilege one side of these tensions over another, neither ignoring member democracy nor business strategy. Brett Fairbarin (1999) argues many agricultural cooperatives in the United States and Canada have historically ignored member democracy, while competing with large investor-

[handwritten: need to be economically sustainable]

oriented firms, by becoming larger bureaucratic organizations in their own right and distant from their local members. Conversely, while many new cooperatives entered the field in the 1970s associated with ecological and organic food movements, many went bankrupt in the 1980s and 1990s due to the failure to form a strong wholesaler organization, a degree of layered bureaucracy (Fairbarin 2004). A cooperative that privileges aspects of democratic collective action, while dismissing business operations that respond to the market, risks its capacity to compete in the marketplace—and ultimately to meet its members' needs. On the other hand, a cooperative that is "all business" and leaves unaddressed basic issues of identity, participation, and meaning, risks losing member loyalty, patronage, and ultimately the ability to meet its members' needs as well. Therefore, cooperative structures designed to meet agriculture of the middle goals must be carefully considered, with these tensions in mind.

Historically, agricultural cooperatives have followed three basic organizational strategies: locals, centralized, and federations—though there is some mixing of types. Each is organized around democratic principles, though each has different authority relationships.

[handwritten: 3 types — 1 locals]

Local Cooperatives

Local cooperatives are the most bottom-up, grassroots of the three structures (see figure 3.1). Local associations generally serve relatively small

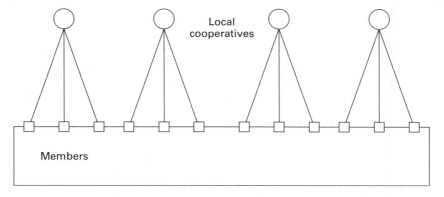

Figure 3.1
Local Cooperatives
Source: Schaars 1971, 51.

numbers of members with similar interests. Typically, a local cooperative may have as few as ten to fifteen members, while others will have as many as five hundred to a thousand (Schaars 1971, 50). Among many possible activities, farmers and ranchers in a cooperative may collectively buy supplies and services, market their output, and/or process and market products. Members usually live within a relatively small area, and many will know each other. Relationships are more informal; mutuality is more easily recognized. While business is the central activity of the organization, member participation (beyond business patronage)—in governance and decision making—can provide the additional benefit of a sense of individual and community efficacy as well as collective identity and meaning. These experiences speak to the issue of collective autonomy and not being cogs in a business machine. (For examples of local cooperatives, see the case studies of Oregon Country Natural Beef and the Thumb Oil Seed Cooperative at ⟨http://www.agofthemiddle.org⟩.)

Democratic control runs from the members as a group to the local association. An elected member board of directors sets policy and may make managerial decisions as well, depending on the degree of complexity of the business. If cooperative business activity is such that it is conducted throughout the year (rather than on a seasonal basis) and business activity has a degree of complexity not easily resolvable with part-time management, then a professional manager is usually hired. With the introduction of a professional manager, the tension between grassroots (member) wisdom and complex (managerial) expertise may become more manifest.

Local cooperatives have the organizational potential to achieve many of the objectives of farmers and ranchers of the middle. Open, transparent, and democratic, they can provide an organizational vehicle for farmer voices and collective mobilization around farmer interests. They are local by definition, and can articulate through action the agendas of socioeconomic and ecological sustainability. As independent businesses, they can represent a decentralized approach to supporting the agendas of a dispersed, independent, and family-based agriculture.

It needs to be mentioned, though, that agriculture of the middle agendas as pursued by many independent locals would lack coordination. Local cooperatives, by definition, would almost certainly be responsive to local agendas as well as local creativity and diversity. They would likely lack coordination and congruency across regional or national operations. Given an intense competitive environment, local co-ops could also find themselves at a competitive disadvantage to firms less socially and ecologically responsible. Larger organizations can be more powerful in the marketplace by creating scale and coordination advantages. Thus, for reasons of scale, scope, and consistency, leaders of the agriculture of the middle initiative suggest that a larger organization, in combination with local operations, may be the most appropriate structure for pursuing interests of the middle (Yee 2004, 2005).

Centralized Cooperatives

Centralized cooperatives are similar to locals in that members belong to a single organization (see figure 3.2). Unlike a local, however, this membership is composed of large numbers (thousands) of farmers spread over broad geographic areas. Typically, the headquarters is far removed from most farm locations. Local facilities exist for servicing members, though these local units are affiliated with the central headquarters organization. They represent local business sites only, for local farmers, and are not themselves cooperatives. The governance system operates from the farmer to the central organization. Members elect a member board of

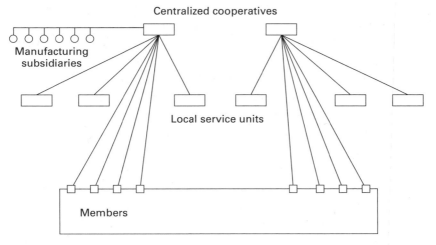

Figure 3.2
Centralized Cooperatives
Source: Schaars 1971, 51.

directors—either directly or through elected delegates. The board then sets policy for the management of the cooperative business. The board is also charged with hiring the cooperative executive officer, who manages from the central headquarters. The management of the local facilities is hired and employed by the management structure of the central organization. Ocean Spray, Southern States, and Dairy Farmers of America are examples of centralized cooperatives.

ex.

Local facilities of centralized associations provide all the services of any local cooperative—for instance, local assembly, grading, packing, shipping, processing, or purchasing. Yet centralized cooperatives have various advantages of scale, scope, and resources that locals do not have. David Cobia (1989) lists these various advantages to include: achieving greater uniformity of products and services regionally, by operating all local units from the center; lowering operating costs with centralized control of the handling and marketing of products; obtaining greater bargaining power in the marketplace, via lower operating costs and the ability to command greater volumes; and achieving the broad ability to adapt locals to rapidly changing economic conditions.

adv. of scale

The nature of centralization is such that while it brings various advantages, it simultaneously brings disadvantages. By definition, decision making is centralized rather than decentralized. Democracy provisions take shape as a democratic bureaucracy rather than a direct participatory democracy (Cobia 1989). The experience of the cooperative as a mechanism for developing and deepening mutual identities—farmers and ranchers of the middle—may all but be eliminated with organizational size and bureaucratic authority flows. Senses of individual or community efficacy, relationship, and meaning can become highly muted. Under such structures members may lose active interest, and potentials for developing exclusionary business rationales (parallel to a short-term return on investment logic) become more likely. It has been the case historically—under the pressure to achieve economic success in competitive markets—that the cooperative membership itself has dismissed its own collective voice in deference to managerial rationalization and authority (Fairbarin 1999, 2004; Hogeland 2005). In turn, the authority of managerial expertise has frequently sacrificed activities that do not make an obvious contribution to the bottom line or short-term business survival. An internal logic parallel to return on investment criteria tends

what is travel

to take precedence, and cooperative operations and member involvements begin to look similar to investor-oriented firms. Under such situations, the probabilities of achieving the goals of agriculture of the middle farmers become more difficult, and consumer trust and connection may . be lost. (For the importance of creating and maintaining connections between farmers and ranchers and food consumers, see chapter 6.)

Federated Cooperative Structures

Local cooperatives may sometimes join together and form a collective or federation of cooperatives (see figure 3.3). In a federation, farmers hold membership in local cooperatives, and in turn, the locals form a cooperative of locals. Locals own the federation and typically provide large proportions of its capital needs. They also elect a board of directors, and that board hires the federation management. Locals continue to operate as local cooperatives with their own respective boards and management, though the federation may sometimes provide—via contract—management of a local. This federated structure can supply the services and most of the advantages of any centralized cooperative. By definition and structure, federations can be as responsive at a local level as any unaffiliated local cooperative.

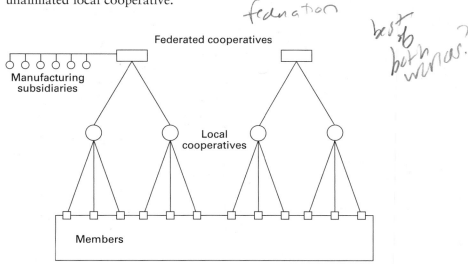

Figure 3.3
Federated Cooperatives
Source: Schaars 1971, 51.

Federations of cooperatives fall in the midrange of organizational dilemmas. They are organizationally complex and bureaucratic, but are structured in a manner to allow for both representative democracy at the federation level and direct participative democracy at the local level. Because the federation is built and controlled in this manner (from the bottom-up), the local members' economic interests may be better expressed through federation membership, while senses of community and identity can more readily be maintained through the direct ties to the local—for example, through patronage, voting, office holding, and informal familiarity (Cobia 1989; Schaars 1971). Federated cooperatives include OFARM, California Pear Growers, and Florida Natural Growers, although CPG and FNG tend to be commodity-agriculture oriented.

Federated cooperatives can address the following insightful observations by Robert Briscoe and Michael Ward (2000, 64)—citing E. F. Schumacher—in *The Competitive Advantages of Cooperatives*:

Whenever one encounters such opposites [as centralization and decentralization], each of them with persuasive arguments in its favor, it is worth looking into the depth of the problem for something more than compromise.... Maybe what we really need is not either-or, but [both] together... we can find ways to enjoy the benefits of size while staying small, we can get the advantages of centralization while remaining decentralized.

Federations are capable of incorporating organizational tensions—particularly centralization/decentralization—such that they are made obvious and therefore more workable, and do not sacrifice one for the other. Furthermore, given the agendas of the agriculture of the middle initiative, and the needs to establish and deepen local as well as collective identities, the concept of "heterarchy" may be useful. Heterarchy refers to organizational arrangements that seek to "coordinate diverse identities without suppressing differences" (Stark 2001a, 2001b). Heterarchical organization has the capacity to deepen the richness of the local while providing for overall coordination. Federations can allow for strengthening local identities—via local cooperation—while offering a central mechanism for general coordination and achieving scale efficiencies.

Leaders of the agriculture of the middle initiative argue in a parallel manner that the interests of the middle might best be pursued with federated-like structures (Yee 2005). Key features to be centralized within a federation might include:

- a common seal that stands behind food products and brands associated with local member cooperatives, and focuses on dimensions that highlight key agriculture of the middle values;
- a third-party certification methodology administered from the federation to bring consistency across the locals and guarantees of food quality to consumers;
- regional and national coordination of cooperative activities and flows of product;
- professional broad-scale marketing and advertising;
- research and education; and other professional support services. (See ⟨http://www.associationoffamilyfarms.org⟩.)

Such federated coordination is likely necessary to enable the marketing of the volume of product associated with "markets of the middle," particularly in a marketplace also occupied by large, investment-oriented, complexly integrated firms. Federations can provide market presence and scale, while supporting local identities and brands.

Nevertheless, any cooperative business must be vigilant about its internal tensions and conflicts (Mooney 2004). A federation of local cooperatives may offer degrees of decentralized decision making and local identity as well as regional or national coordination. Yet the same dynamic trade-offs will exist within a federation as in a centralized cooperative. The tension between managerial expertise and grassroots wisdom, between demands for business efficiency and democratic participation, will likely be present. Mooney (2004) suggests making these tensions explicit, and then planning for them, as a way to help keep federated cooperatives embedded within member prerogatives and local needs. Hiding from these tensions with poorly designed or maintained participation structures will likely result in cooperative failure as a democratic organization. Under such conditions, consumer connections with agriculture of the middle values are likely to be compromised as well.

Conclusion

The primary message of this chapter is to consider various socioeconomic, social-psychological, and cooperative organizational perspectives to address agendas for renewing an agriculture of the middle. Much of the chapter's conceptualization has been conditioned by a view that investor-oriented firms will relentlessly pursue vehicles for investment into profitable activities. Given the pursuit of profits—and the pursuit of market power to realize profits—the processes of agrifood industrialization and corporate conglomeration are likely to continue. Farmers and ranchers of the middle then have the task to find a place between increasingly globalized food systems that reward commodification, industrialization, and scale, and much smaller local food systems that reward food product differentiation, direct marketing, and local identities.

From the perspectives of high modernity theories (Giddens 1990, 1991; Gertler 2004; Bauman 2000; Beck 1992; Gray 2000), increasing numbers of consumers want products that connect with personal, community, and ecological sustainability. In addition, new social movement and collective action theories focus on a considerable and generalized dissatisfaction with anonymous power along with the lack of accountability of big businesses (Buechler 1995; Larana et al. 1994; Pichardo 1997). These issues are reflected in growing demands for self-determination, authentic discourse, and transparency in the marketplace. Farming and food systems that honestly engage these issues and faithfully represent the core values of sustainability—the permanent significance of economic, ecological, and community well-being—are likely to have advantages in earning the loyalty of such consumers.

Cooperatives have their own organizational advantages in terms of democratic structure, transparency, and service. As democratic participatory organizations, they provide a forum for farmers and ranchers to express their own individual interests as well as a mechanism for building a collective identity. Cooperatives have also been effective historically in organizing farmers for power, particularly in opposition to monopoly and oligopoly interests. This will be extremely important in an ever-competitive marketplace. Potential earnings, and even agriculture of the middle economic successes, will entice the interests of investor-oriented firms to compete for products and markets. These competitive pulls can shift cooperative purposes away from their original bases toward a more short-term return on investment logic.

This chapter suggests a federated cooperative structure may be the most appropriate for agriculture of the middle agendas. The structure offers an approach to heterarchy—coordinating and enriching diversity—based on an overarching representative democracy designed to coordinate participatory locals. Historically, federated cooperatives have been able to compete with much larger organizations in the marketplace. They also provide for a system of transparency as well as accountability—crucial qualities for an increasing number of contemporary food consumers. Still, those utilizing cooperative models will need to be vigilant about the various tensions inherent in these organizations, and the dynamics that can shift and privilege some benefits at the expense of others.

Note

The views expressed in this chapter are solely those of the authors, and do not reflect the positions or policies of any associated department, school, or administration.

References

Bauman, Zygmunt. 2000. *Liquid Modernity*. Oxford: Polity Press.

Beck, Ulrich. 1992. *Risk Society: Towards a New Modernity*. London: Sage Publications.

Briscoe, Robert, and Michael Ward. 2000. *The Competitive Advantages of Cooperatives*. Cork: Centre for Co-operative Studies, National University of Ireland.

Buechler, Steven. 1995. New Social Movement. *The Sociological Quarterly* 36 (3): 441–464.

Cobia, David. 1989. *Cooperatives in Agriculture*. Englewood Cliffs, NJ: Prentice Hall.

Craig, John. 1993. *The Nature of Cooperation*. Montreal: Black Rose Books.

Dunn, John. 1988. Basic Cooperative Principles and Their Relationship to Selected Practices. *Journal of Agricultural Cooperation* 3:83–93.

Fairbarin, Brett. 1999. The Historic Basis and Need for Cooperatives: Can Cooperatives Maintain Their Historic Character and Compete in the Changed Environment? In *Proceedings of Farmer Cooperatives in the 21st Century*. Ames: Department of Agriculture Economics, Iowa State University.

Fairbarin, Brett. 2004. History of Cooperatives. In *Cooperatives and Local Development*, ed. Christopher D. Merrett and Norman Walzer, 23–51. Armonk, NY: M. E. Sharpe Press.

Gertler, Michael. 2004. Synergy and Strategic Advantage: Cooperatives and Sustainable Development. *Journal of Cooperatives* 19:32–46.

Giddens, Anthony. 1990. *The Consequences of Modernity*. Cambridge, UK: Polity Press.

Giddens, Anthony. 1991. *Modernity and Self-Identity: Self and Society in the Late Modern Age*. Cambridge, UK: Polity Press.

Gray, Thomas W. 2000. High Modernity, New Agriculture, and Agricultural Cooperatives. *Journal of Cooperatives* 15:63–73.

Gray, Thomas W., William Heffernan, and Mary Hendrickson. 2001. Agricultural Cooperatives and Dilemmas of Survival. *Journal of Rural Cooperation* 29:2:167–192.

Hogeland, Julie. 2005. New Generation Co-op, Limited Liability Corporation, Value-Added, Demutualized: What Is Still "Cooperative" about American Agricultural Cooperatives? Paper presented at the twenty-first International Cooperatives Research Conference, Cork, Ireland, August 12.

Johnston, Hank, Enrique Larana, and Joseph R. Gusfield. 1994. Identities, Grievances, and New Social Movements. In *New Social Movements: From Ideology to Identity*, ed. Enrique Larana, Hank Johnston, and Joseph R. Gusfield, 3–35. Philadelphia: Temple University Press.

Larana, Enrique, Hank Johnston, and Joseph R. Gusfield, eds. 1994. *New Social Movements: From Ideology to Identity*. Philadelphia: Temple University Press.

Lyson, Thomas A. 2004. *Civic Agriculture: Reconnecting Farm, Food, and Community*. Medford, MA: Tufts University Press.

Lyson, Thomas A., and Joanna Green. 1999. The Agriculture Marketscape: A Framework for Sustaining Agriculture and Communities in the Northeast. *Journal of Sustainable Agriculture* 15:108–113.

Melucci, Alberto. 1988. Getting Involved: Identity and Mobilization in Social Movement. *International Social Movement Research* 1:329–348.

Melucci, Alberto. 1996. *Challenging Codes: Collective Action in the Information Age*. Cambridge: Cambridge University Press.

Miller, D. 1995. Consumption as the Vanguard of History. In *Acknowledging Consumption*, ed. D. Miller, 1–57. New York: Routlege.

Mooney, Patrick. 2004. Democratizing Rural Economy: Institutional Friction, Sustainable Struggle, and the Cooperative Movement. *Rural Sociology* 69 (1): 76–98.

Pichardo, Nelson A. 1997. New Social Movements: A Critical Review. *Annual Review of Sociology* 23 (1): 411–430.

Schaars, Marvin. 1971. Cooperatives, Principles, and Practices. Madison, WI: University Center for Cooperatives.

Schumacher, E. F. 1974. *Small Is Beautiful: A Study in Economics as if People Mattered*. London: Abacus.

Sexton, Richard. 1997. The Role of Cooperatives in Increasingly Concentrated Agricultural Markets. In *Cooperatives: Their Importance in the Future Food and Agricultural System*, ed. Michael Cook, Randall Torgerson, Thomas Sporleder, and Donald Padberg, 31–47. Washington, DC: National Council of Farmer Cooperatives.

Stark, David. 2001a. Ambiguous Assets for Uncertain Environments: Heterarchy in Post-Socialist Firms. In *The 21st Century Firm: Changing Economic Organizations in an International Perspective*, ed. Paul DiMaggio, 69–104. Princeton, NJ: Princeton University Press.

Stark, David. 2001b. Heterarchy, Exploiting Ambiguity, and Organizing Diversity. *Brazilian Journal of Political Economy* 21 (1): 2–79.

Stevenson, G. W. 1998. Agrifood Systems for Competent, Ordinary People. *Agriculture and Human Values* 15 (3): 199–207.

Thu, Kendall, and E. Paul Durrenberger, eds. 1997. *Pigs, Profits, and Rural Communities*. Albany, NY: University of New York Press.

U.S. Department of Agriculture (USDA). 2004. *Farmer Cooperative Statistics, 2002*. Rural Business-Cooperative Service, service report 62. Washington, DC: Rural Development-Cooperative Programs.

Yee, Lawrence. 2004. Strategy for an Agriculture of the Middle. Paper presented at the Agriculture of the Middle Task Force meeting, Racine, WI, July.

Yee, Lawrence. 2005. The Association of Family Farms. Davis: University of California.

4

Is Relationship Marketing an Alternative to the Corporatization of Organics? A Case Study of OFARM

Amy Guptill and Rick Welsh

"relationship marketing"
+ flight from farming

In the words of Linda Lobao and Katherine Meyer (2001, 103), the twentieth century has seen "the national abandonment of farming as a livelihood strategy." They point out that farmers were about one-third of the U.S. population in 1900 but less than 2 percent by 2000 (103). In the current era, Census of Agriculture statistics reveal that 90 percent of the income accruing to small farms (less than $250,000 annually, or 98 percent of all farms) comes from off the farm. In light of global-scale agricultural markets and international challenges to agricultural subsidies, some, like economist Steven Blank (1998), have predicted "the end of agriculture in the American portfolio." Creative solutions are needed to revitalize productive landscapes in the rural United States, and those solutions must build sustainably on the natural, cultural, and social resources of contemporary U.S. farmers. As Jan Douwe van der Ploeg (2000) reminds us, the economic viability of farms is still a key component of rural development.

Rural sociologists and agricultural economists have traced the cost-price squeeze undermining U.S. agriculture to a production-marketing-processing system that maximizes the profit accruing to large agricultural corporations at the expense of producers. This system is increasingly *vertical integration of production* dependent on contracting to maintain vertically integrated production chains. There are several competing explanations of why U.S. agriculture has changed in this way (Welsh 1996). It is clear, however, that such changes could foster opportunities for farmers to organize themselves for collective bargaining. Recent research shows that farmers marketing together tend to gain more favorable prices and terms of trade than they can on their own (chapter 10, this volume; Levins 2001).

Farmers' right to bargain collectively was established by the Capper-Volsted Act of 1922 and affirmed in the Agricultural Fair Trade Practices

Act of 1967. Essentially, Capper-Volsted exempts farmers from the antitrust legislation that prevents other kinds of producers from sharing production and price information. To qualify for privileges under Capper-Volsted, farmers' organizations such as cooperatives, limited liability corporations, or other entities must be made up entirely of farmers engaged in the production of agricultural products, run with a one-member-one-vote system, and not paying more than 8 percent in dividends on capital. These organizations can handle the products of nonmembers as long as that produce does not exceed the amount of member produce handled over the course of the year (Volkin 1995).

Analysts predicted that as vertical integration and contracting took hold, cooperative agricultural bargaining associations as envisioned in the Capper-Volsted Act would become more numerous and prominent. Yet contrary to expectations, as contract growing has become a larger part of the market, the number of active bargaining associations has decreased. In 1978, there were sixty-seven bargaining associations spanning thirteen states, but by 1997, only nineteen associations existed in nine states (chapter 10, this volume). Analysts tend to blame this trend on the lack of legal protections for grower organizing efforts (ibid.). As analysts have noted, the federal Agricultural Fair Trade Practices Act of 1967 affirms the right of farmers to bargain collectively but fails to provide powerful enforcement mechanisms to make such abstract rights a reality. A number of states have adopted laws that are stronger than the basic federal rules, and this has been shown to make a difference in the experiences and long-term viability of farming in different states (ibid.). As the farm crisis of the 1980s extended into the following three decades, the rise of organic agriculture seemed to offer a way to sustain independent production. While many organic farmers are new to agriculture, the growth of the organic market has drawn in conventional growers, particularly in field crops where organic production is similarly mechanized and marketing is similarly commodified.

Organic Farmers' Association for Relationship Marketing

The founding of OFARM, and most of its member organizations, coincided with a boom in the sale of organic products beginning in the late 1990s. Despite the flourishing demand, organic grain growers were finding themselves disadvantaged in marketing due to the lack of reliable marketing information. Discussions about establishing an umbrella group for organic producers began in 1998, and OFARM was officially founded in early 2001. OFARM's mission is to "coordinate efforts of producer marketing groups to benefit and sustain organic producers" (⟨http://www.ofarm.org/⟩). It currently includes eight member organizations, and as described on its Web site, represents "the largest single organized block of production in North America covering producers in eighteen states and Ontario."

In legal terms, OFARM is a *marketing agency in common*, comprised of organizations that satisfy the Capper-Volsted requirements of farmer ownership and control. OFARM could market members' grain, but the board has decided that individual organizations will continue to market grain while OFARM provides an overall umbrella. Some of the efforts of OFARM include exchanging pricing and marketing information, educating policymakers, collectively bargaining with the buyers of members' products, and assisting farmers in adopting new crops and agronomic practices in their crop rotations. Currently, OFARM members interact in monthly conference calls for marketers, monthly conference calls of the board of directors, a fall board meeting, and a January annual meeting open to any farmer member of any organization.

OFARM is an interesting case study for at least two reasons. First, at this point, the organization represents primarily field crop producers in the Midwest—a sector that has not received extensive attention in the literature on organics but has been shown to be quite different from fruits and vegetables in terms of historical patterns as well as structure (Hall and Mogyrody 2001). In this way, it offers new empirical evidence to further articulate the character and scope of the "conventionalization" of organics (Guthman 2004), particularly the impact of agribusiness on shaping the possibilities for growers within the industry.

Second, as giant food corporations increasingly buy in to the organic sector, the successes and disappointments of OFARM will both reveal and shape the extent to which organics will continue to offer North American field crop farmers a viable alternative to corporate-controlled conventional markets. New inquiry is needed into the prospects and possibilities associated with preserving or generating family-scale production systems that enable the independence of small-scale production and stabilize the livelihood of midsize economic units. In manufacturing, researchers have noted that concentration and vertical integration throughout the twentieth century led to, in some places, the resurgence of a small-scale, craft-oriented manufacturing sector specializing in high-quality, reliable, and flexible production (Piore and Sabel 1984). Producers in these cases organized to balance cooperation with competition and to compete on the basis of quality rather than rock-bottom price. As the huge industrial scale of agriculture follows the same path, analysts wonder if producers in the food system will create similar collaborative solutions to a difficult marketing situation (Lyson and Geisler 1992). Organic agriculture, as a differentiated, high-value product within a larger industrialized sector, seems a reasonable analogue to that trend.

This chapter reports the results of a yearlong case study of OFARM and links these preliminary findings to those two broader concerns in the sociology of agriculture literature. The purpose of the case study was to investigate the structure and functioning of OFARM, and preliminarily assess the capacity of this organizational model to sustain the independence and economic viability that drew its member farmers into organics. The questions driving the case study are basic: What are OFARM's goals? How well is it working? How can OFARM's successes and challenges be understood? Sociologically speaking, those empirical questions led to two others: What information does OFARM add to the conventionalization debate? And in what ways does OFARM offer a powerful organizational model for regenerating a stable agricultural middle? The chapter reviews the conventionalization debate along with recent findings about industrial districts, clusters, and networks. Following that, we explain our research methodology as well as interview and observational findings about OFARM's goals, key strengths, and current challenges. The concluding discussion links OFARM's findings to those two larger themes, and outlines future research questions about OFARM and similar efforts.

The Conventionalization Debate

In the past decade, the explosive growth of organics attracted the attention of scholars and corporations alike. Agrifood studies have noted the emergence of commodity systems in organics alongside the "old *[handwritten: debate over the conventionalization [t] organics]*

guard" direct-market/craft-growing systems that were, until recently, the hallmark of the movement (Guthman 2004). Within this literature is a productive debate about the extent to which we are seeing the conventionalization of organics—compromised organic standards mapped on to an otherwise-conventional (industrial) commodity system, without the ecological integrity and multidimensional progressive values of the original movement. Daniel Buck and colleagues (1997) were among the first to highlight the issue when their analysis of the organic commodity system of northern California revealed that agribusiness capital is "finding ways to reshape the particularities of organic agriculture to its own advantage" by getting involved in "high-value and high-turnover crops," inputs and postharvest processing, and distribution and retailing (ibid., 17).

Brad Coombes and Hugh Campbell (1998) contested the conventionalization thesis, arguing instead that a conventionalized system could exist side by side with the more alternative system, as natural and social barriers to commodification. Research in Ontario by Alan Hall and Veronika Mogyrody (2001) found that organic fruit and vegetable growers were not conventionalizing in the terms laid out by Buck and colleagues (1997), nor was the sector bifurcating into industrialized and localized poles, as Coombes and Campbell observed (1998). In field crops, however, Hall and Mogyrody noted that organic producers were similar to their conventional counterparts in terms of highly capitalized, mechanized, and specialized production for export. Yet they failed to find evidence that the entry of conventional farmers into organics was accelerating or creating undue competition for existing growers. Overall, Hall and Mogyrody emphasize the limits of the conventional-alternative duality to explain the diversity of strategies and values among organic producers in Ontario, thereby suggesting that we should recast the terms of the conventionalization debate.

Julie Guthman's "rejoinder to the 'conventionalisation' debate" brings more data to bear on the argument that conventionalization is happening and can be traced to corporate encroachment. She writes that "agribusiness involvement does more than create a soft path of sustainability"; rather, it sets conditions that "undermine the ability of even the most committed producers to practice a purely alternative form of organic farming" by "[driving] wider processes of agro-industrialization"

(Guthman 2004, 301–302). Agribusiness does this, according to Guthman, by trying to influence the setting of organic standards, appropriating the most profitable organic products (like salad mix), and indirectly forcing smaller growers to reconcile the new competitive pressures with their own agroecological commitments through intensification of production. In all of these areas, Guthman essentially traces conventionalization to the corporatization of the commodity system in a way that shapes the experiences of all growers and producers in the sector.

Most of this debate has addressed conventionalization from consumers' or activists' concerns in terms of whether the conditions of production, processing, and distribution reflect the progressive values that lead some people to support them. From that perspective, organic field crop producers who limit their rotations to take advantage of hot prices in particular commodities (as Hall and Mogyrody found) are exploiting the limits of auditable organic standards to turn natural capital into personal wealth—a very conventional approach to farming. As Guthman (2004) points out, standards themselves are inherently insufficient to guarantee that the producers have deep commitments to the values that many consumers associate with the organic label. In examining producers, though, Hall and Mogyrody (2001) found that the ideological orientations of organic farmers are complex and not easy to infer from the grower's background, experience with organics, or even current marketing practices.

As explained below, the founders of OFARM see the conventionalization of organics in terms of the falling profits and loss of control that their constituency experienced as conventional farmers in the 1970s and 1980s. Everard Smith and Terry Marsden argue that this facet of conventionalization has already begun in the United Kingdom as a slowdown in the growth of the organics market and the prominent role of supermarkets in recent growth in consumption is precipitating "the traditional farm-gate price-squeeze, so long an important feature of conventional agriculture" (2004, 345). This cost-price squeeze comes from both horizontal integration (larger-scale processing in a single sector), vertical integration (contracting), and now strategic alliances among agrifood corporations that limit farmers' production and marketing options (Heffernan 2000; Heffernan et al. 2001). In this way, the problems faced by farmers in the system converges with Guthman's (2004) elaboration of the conventionalization thesis, in that it comes down to the role of large corporations shaping the entire commodity system to their advantage.

The production and processing stages of corporate-controlled agricultural commodity chains has been decidedly industrial with the attendant social and environmental costs (Friedland 2004; Hinrichs and Welsh 2003; Welsh 1996). It's important to recognize that for most of the producers in the OFARM network, however, farming organically is not the alternative to farming conventionally but rather the alternative to not farming at all. Their switch to organics has as much to do with higher prices and greater autonomy than with environmental and social stewardship. In this sense, the producers that OFARM represents are less like organic market gardeners and more like the medium-scale manufacturers displaced in the latter part of the twentieth century through corporate consolidation. With this analogy, OFARM itself is reminiscent of the networks of small manufacturing plants and other firms that emerged in particular places around the world: a network of specialized, quality-focused producers balancing cooperation and competition within a framework of trust. We begin to address our goal of understanding the emergence of efforts like OFARM and predicting their outcomes by examining the literature on small firm networks in other economic sectors.

The Second Agricultural Divide?

The intractable decline of family farming throughout the twentieth century does not reflect the corporate takeover of farming per se (Lobao and Meyer 2001). Rather, corporations have sought to control distribution and processing in order to capture more of the value in the chain (Heffernan 2000; Hendrickson et al. 2001; Levins 2001). Some family farms have persisted by subsidizing their farms with off-farm work, absorbing the risks associated with farming, and deploying local knowledge to enhance production (Lyson 2004). As rising costs of production and falling prices have continued to squeeze producers, some have sought to build alternative forms of production and marketing outside the corporate-controlled system (Lyson 2004). Organics is one prominent example. While the shape of trends in agriculture is clearly distinct from other areas of the economy, the patterns of corporate control in agriculture have some connection to the postindustrial trends in the manufacturing and technology sectors (Lyson and Geisler 1992).

In the mid-1980s, Michael Piore and Charles Sabel (1984) touched off an explosion of research on the functions of and interrelationships among small and midsize enterprises in the era of the megacorporation. They noted the emergence of a cluster of small-scale specialty manufacturing plants in the "Third Italy" that thrived in the wake of consolidation in manufacturing. For Piore and Sabel, these firms represented the reemergence of a craft-based knowledge-intensive form of production that they termed *flexible specialization*. With highly skilled employees, flexible technologies, and dense interfirm networks of subcontracting and information sharing, these firms created a favorable role for themselves within the new economy. By balancing cooperation and competition, and competing on quality rather than price, Piore and Sabel argued, these firms supported stable livelihoods and the economic vitality of the broader region. Research following their book, *The Second Industrial Divide*, has sought to elucidate the benefits of networking relationships for small firms and the extent to which these networks benefit localities.

In the first category, many studies have underscored how networks act as a shared resource that facilitates organizational learning. Early analyses have stressed the importance of trust in maintaining the interfirm relationships that enable flexible specialization (Perrow 1993; Sabel 1992). In a recent example, Boris Braun and colleagues (2002) compared small firm networks in the mechanical engineering/metalworking industry in Germany with those of its counterpart in Japan, and found that German firms enjoyed more "strategic freedom" (flexibility and autonomy) because they more often had distinct identities in the final market and a less hierarchically structured network. Agriculturally, Lyson and colleagues (2000) found that dairy farmers with community ties show a stronger economic performance than those who lack such connections. While imprecision in documenting the structure, function, and geography of firm networks has hampered robust comparison (Staber 2001) among regions, industries, or phases of the industry life cycle, overall the notion that networks of small and medium enterprises support innovation and learning that enhances firm and regional competitiveness is well established in the economic geography and regional studies literatures. Do these networks help sustain small firms? Systematic evidence is lacking, but the findings of Braun and colleagues (2002), John Britton (2003), Andy Cumbers and colleagues (2003), and Martin Kenney and Donald Patton (2005), among others, suggest that how a firm is situated within the commodity chain is ultimately as important in sustaining small firms as their participation in local innovative networks.

Do benefits to smaller firms mean benefits for localities? While this is less well specified than the boost to innovation that networks foster, there is some strong evidence from both manufacturing and agricultural sectors that in general, small and medium firms strengthen local economies. In manufacturing and high-tech industries some have found geographic clustering in some cases, which tends to benefit local economies (Pietrykowski 1999; Kenney and Patton 2005), leading some observers to predict the "re-regionalization" of manufacturing. Guptill (1998) and Thomas Lyson and Guptill (2004) find similar effects in agriculture, in that conventional industrialized production and alternative "civic" forms become tied to distinct regional variations. Further research provides strong support for the idea that places characterized by small and mid-size firms make a contribution to local well-being net of confounding variables (Tolbert et al. 1998; Lyson et al. 2001). Gary Green (1994) suggests that midsize firms carry more local benefits than small ones, including a greater capacity for innovation. While smaller firms are clearly more beneficial than large ones, the exact mechanisms of that benefit and the links to firm networks remain to be specified.

To what extent, then, is this line of research applicable to agriculture in general and OFARM in particular? While it is too soon to make clear comparisons or confident predictions, the literature on firm networks does offer a point of departure for making sense of OFARM's goals and structure. First, it is empirically and intuitively defensible to say that networks do help these firms access information, pool resources for research and development, and in some cases, exercise some governance over the markets in which they participate. For that reason, it's reasonable to hypothesize that a group like OFARM will produce tangible benefits for its members. Second, this literature emphasizes the importance of trust, transparency, and the autonomy of each firm within the network. That suggests that the format OFARM has chosen—to be a marketing agency in common rather than a large cooperative, will prove to be a crucial factor in its success. Third, as explained below, OFARM's key goal of capturing more *power* (not just value) in the organic supply chain reso-nates with findings in the industrial literature about how the benefits of firm networks articulate with broader issues of market power. With these issues in mind, we conducted a yearlong case study of OFARM to investigate its goals, structure, and operation.

Research Methods

We employed an inductive qualitative approach to investigating the structure, function, and effectiveness of OFARM (Strauss and Corbin 1998). To gather data, we observed the work of the organization during three conference calls and two daylong face-to-face meetings; conducted seven individual interviews with members of the board, the marketing directors, and OFARM staff; and reviewed the organization's back-ground documents and newsletters. The qualitative methods used in the study—interviewing and observation—draw on a "grounded theory" approach, in which researchers probe emerging questions and test out tentative conclusions with further research. Researchers stop collecting data when the findings reach a point of "saturation," in which new data tend to confirm working conclusions rather than reveal new critical dimensions (Strauss and Corbin 1998). In this project, the notes from the conference calls and fall 2004 board meeting informed the construc-tion of the semistructured interview guide. The interview results, in turn, guided the analysis of the field notes taken at the annual meeting in Jan-uary 2005.

Most of the interviews were conducted over the telephone and took thirty to forty minutes to complete. Most interviewees represent member organizations as OFARM board members, marketing agents, or both. One was an OFARM staff member. They were asked for background in-formation on their own organizations, their views on OFARM's strengths and challenges, and their perceptions of the alignment of OFARM's goals with those of member organizations. Two member organizations were not represented in the interviews; one representative declined to participate, and another was unreachable. Nevertheless, the six com-pleted interviews produced remarkably similar findings, which were rein-forced by notes from marketing calls and meetings. For this reason, we are confident in drawing these conclusions about OFARM's functioning and effectiveness. To preserve confidentiality with a small sample size, we have combined quotations from interviews and observations in our discussion of the results, and referred to all speakers vaguely as "members." Drafts of this and other writings from the project have been reviewed by OFARM prior to their release.

Results

The results of our case study suggest five preliminary points:

1. OFARM's goal is to prevent the conventionalization of organics by amassing and exercising economic power to create an orderly market in organics. This goes beyond getting more of the final food dollar, but instead involves collaborating with other like-minded actors in the food chain to sustain the industry.

2. The main benefit of OFARM that members describe is the exchange of information that enables participants to "know the value of their product," and as a result, negotiate better terms with buyers. More minor, but significant benefits include mentorship and camaraderie.

3. Factors contributing to OFARM's success include the trust developed among members, the flexibility afforded by the marketing-agency-in-common business model, and the seemingly inexhaustible demand for organic field crops.

4. Paradoxically, organizing nationally has led in some cases to more regionally rationalized markets.

5. The challenges OFARM faces come largely from deep-pocketed food corporations undermining farmers' connections to organic marketing organizations, the need to recruit more members, and the constraints imposed by the Capper-Volsted legislation.

We discuss each of these points in turn below.

OFARM's Goals

In short, OFARM's main goal is to prevent the conventionalization of organics. As one member explains, "We want a fair and decent price for the grower, [and] consistency in the market. We want to unite the organic community more that it is, so that they don't become a product of the conventional market. We don't want to conventionalize the organic market. We don't want to let it be taken over by the big conglomerates where we're slave labor out here. We want to maintain smaller farms." Another member concurs: "On the conventional side we've seen what's happened—low prices for farmers and a market controlled by the bigger buyers, like Cargill and the rest. Something would have to be done. We talked about the price of production and what you need to make it workable. We didn't want to get involved in government programs."

So while conventionalization, in this formulation, is expressed in terms of the farm-gate price, OFARM's strategy for maintaining the viability of organic field crops is to exercise more power over the market. As one member puts it, "We didn't know what we would gain, but we had to do something to stabilize the prices. There were a lot of spikes in prices. We wanted to establish a market and establish prices." Another member notes that "we all get on the same page and put the same story out there, and regulate prices to have good orderly marketing."

For these marketers, then, OFARM is understood as not only a useful strategy but a necessary one. With something as simple as a monthly conference call to discuss weather, crops, and market conditions, OFARM hopes to change the balance of power in the growing organic commodity market.

While "stabilizing prices and keeping them up" is the shared goal of interviewees, it was clear from the interview and observation data that price considerations were tempered by a desire to stabilize the industry. Some worried that high organic commodity prices were "stifling" the animal product commodity systems they participated in, and all were clearly concerned about supporting the industry as a whole. In this sense, their desire for power in the industry involves accepting responsibility for planning an orderly growth of quantities and prices.

OFARM's Key Strengths

OFARM has maintained a committed core group of members whose appreciation for OFARM continues to grow. All interviewees voiced strong support for OFARM as a critically important part of their own organization's mission. The most frequently emphasized benefit is the marketing information that comes through OFARM. Several interviewees used the same verbatim phrase to summarize the benefit of OFARM: "knowing the value of our product." They learn the value of their product through the prices that buyers have paid others, but also by information about weather and plantings that help them anticipate supply and quality. As one explains:

Our member producers benefit from the marketing knowledge I gain through OFARM—knowledge about a nationwide area. We know the quantities and qualities and prices out there. Our farmers feel protected under the umbrella; they're not going out there all by themselves.

We know what the markets are going to be, in advance of what they are.

From OFARM, we can really know what our product is worth, through information on weather and market trends—that the value of the product is now.

One member points out that having such information is not unusual: "Cargill has satellites out there, and that's how they became successful. Now we have that too." Similarly, another comments that "with OFARM, our members now have a sense of being part of a larger organization, and a sense of marketing power. Farmers haven't had that sense of marketing power."   *a sense of marketing power*

All interviewees emphasized that the information coming through OFARM was indispensable to their work, and thus only strengthened their ability to meet their obligations to their members.

In addition to the valuable market-related information, some members note how OFARM provides opportunities for mentorship, mutual support, and joint promotion. Several of those relatively new to national-scale organic marketing commented that OFARM also offers them relationships with more experienced growers and marketers who they can and do call for advice. Similarly, some said that they find it reassuring that, in the words of one member, "there are others trying to accomplish the same thing you are—stabilize prices and keep them up." Others mention that promotion of OFARM and member organizations goes together. "Since we've been marketing, the whole concept has gained visibility," remarks one member, "OFARM is pretty visible out there. They always try to get some booth space where organic producers are going to be. That and the Web site has helped our visibility, and I get numerous calls from producers saying they want OFARM to market their grain."

Overall, then, the relationships formed through OFARM go beyond the direct exchange of information to other forms of mutual assistance. These relationships have enabled OFARM to withstand an initial hostile reaction by buyers and establish some basic conditions for "orderly marketing," with stable, fair prices for organic commodities. In interviews and at meetings, members frequently recount early attempts by buyers to claim superior marketing information. In one example, a buyer offers a low price for a commodity, alleging that a large harvest is under way in a different region of the country. "So I got on the phone," this member explains, "and I find out it isn't true." This marketer held out for the higher price, which he eventually received. One member reports taking the connection a step further, by refusing to do business with one buyer until that business paid an overdue bill to another OFARM member. At this point, several members report, buyers facing short supplies are coming to appreciate the ability of OFARM members to direct them to other sources, fill large orders by collaborating with other members, or offer multiyear contracts. "OFARM has a reputation now for knowing what will happen," one member says. "Everything we've been saying about feed grains running out has happened." Moreover, the network that OFARM represents has some benefits for buyers. "If we get a call and we are unable to fulfill an order," a member notes, "then we are comfortable calling other members and having them fill it—it makes each of us a more reliable supplier, working as a group rather than as individuals. There's a benefit for buyers to connect into the OFARM group to have possible relationships with a source."

As a result, some members believe, a lot of buyers are starting to appreciate what OFARM contributes to the value chain as a whole. For example, one member comments: "Initially, we were somewhat a threat to the buyers, but they now know that we can be a reliable supplier and we'll give them a fair price. Initially, they were constantly looking for the lowest price. That's still the case, but not in all cases. A lot of buyers are looking for supply at a reasonable price. That's what we sell." It is thus safe to say that OFARM's main goal of collaborating to exercise more power in the market is already being met.

Factors Contributing to OFARM's Initial Success
OFARM has successfully galvanized a committed core group of members that trust one another and the OFARM umbrella. OFARM enables the exchange of valuable information, members emphasize, because they trust one another. As one member explains, "OFARM provides a comfort zone to share that information. We know that none of us is going

to encroach on another organization's market. OFARM gives us a neutral place to share information, and we respect one another." *face-to-face*

OFARM members make it a point to have face-to-face meetings twice
a year rather than depend entirely on telecommunications, because "the
group tends to gel better" and generate the trust on which OFARM
depends. As a result, no one during the interviews and observations
voiced suspicions about untruthfulness on the part of others. And when
interviewees were asked about the weaknesses or challenges of OFARM,
no one mentioned bad faith among members as a problem.

To further probe the degree of trust within OFARM, member interviewees were asked to imagine the conditions that would lead them
to pull out of OFARM, and all concurred that the organization would
have to completely break down. They would only leave, members say,
"if it can't function," "if people stopped being truthful about what was
going on out there and tried to take advantage of others," and "if we
started to see organizations working against each other." No one can
envision a situation in which OFARM continues to function but going
it alone would better serve their organization's needs. In this sense,
the respondents do not perceive any competition between their service
to their own members and their membership in OFARM as it now
functions.

While it is probably impossible to pinpoint where this trust came from,
several members highlight the important part that the Institute for Rural
America, an organization affiliated with the National Farmers Organization, played in putting OFARM on to a strong footing. The National
Farmers Organization helped OFARM get started by facilitating an
early meeting, identifying potential member organizations, connecting
OFARM organizers with people familiar with the marketing-agency-in-
common structure as an organizational umbrella, and providing initial
in-kind resources like legal counsel, office support, and conference-
calling facilities. Some of OFARM's initial success must be attributed to
the catalyzing role that the National Farmers Organization and the Institute for Rural America played in making the needed connections.

The flexibility afforded by the marketing-agency-in-common structure
has turned out to be a key component of OFARM's initial success. It has
enabled participants to begin with an activity entailing little resource

investment (monthly conference calls), and then create the connections that enable more high-stakes investments (perhaps joint shipping, marketing, and input purchasing) as trust and experience build. Meanwhile, because member organizations continue to do their own marketing, they retain full autonomy as organizations. In the words of one, "Every time we meet we gain new appreciation for what OFARM does for us. We all see the benefits of working together." OFARM was able to start small and build on a social infrastructure of trust. Contrariwise, projects that begin with a heavy financial investment might enjoy less time to make thoughtful, deliberative decisions that maintain connections to the group.

Finally, OFARM as an organization has benefited from the seemingly inexhaustible demand for organic goods. The ever-increasing demand has enabled the organization to set price floors, realize an early tangible success, and make longer-range plans. While several interviewees spoke of respecting one another's existing marketing relationships, the fact that demand heavily outstrips supply makes that an easy courtesy to extend. It also instantly affords them power vis-à-vis buyers without having the vast majority of the organic supply represented under the umbrella. As one member puts it, "I've been telling my buyers to make a contract, and they haven't been interested, but now they're coming to me wanting three-year forward contracts." The member adds, "We are writing the contracts rather than considering the buyers' contracts." While this situation won't last forever, it affords OFARM an incubation period in which to build the relationships both within and beyond the organization to withstand more challenging times.

Regionalization of Organic Commodity Markets

OFARM's goal of creating a stable, orderly market for organic commodities is most frequently expressed in terms of price, but members are also finding that OFARM helps bring rationality to the geographic movement of organic products. First, with a bird's-eye view of the available supplies, members are more often able to connect to markets closer to home, sometimes even encouraging buyers to instead purchase from a closer source. "I've been looking to ship our grain much closer," remarks one member. "Now I can go hundreds rather than thousands of miles. A little better price for the producers, less freight, and not using so much energy. OFARM didn't influence or detract from that, but it supported the idea that there is a market out there." Another elaborates, maintaining that "our long-range goal is to have production organized that meets the needs of markets in our state and the region.... Through OFARM, we actually get to do a supply and demand situation."

Second, one regional farmers' group has formed specifically under the OFARM umbrella and has benefited from the early access to market intelligence as well as technical assistance from OFARM. Third, OFARM seeks to recruit more growers to organics and more members to OFARM in order to better supply the needs of the growing market. It discusses, for example, focusing on growers that are close to growing markets. In this way, a national-scale network is, somewhat paradoxically, supporting the rational regionalization of marketing.

OFARM's Current Challenges

While the organic boom has several important benefits for OFARM, it also has a drawback: it encourages farmers within these organizations to market to the deep-pocketed corporations now getting involved in organics rather than through cooperatives. While OFARM's visibility has helped to recruit members to these organizations, the boom in organics means that buyers are knocking on farmers' doors. "The buyer side isn't the problem," one member explains, "the problem is farmer loyalty." This member adds that "the demand dynamics have changed. It's a paradigm shift. Before we had to push organics, explain it and educate people. But now there's big money involved, and we're being pulled. Demand is outstripping supply. For farmers, it creates a lot of people knocking on their door and it undermines loyalty." As another member reports, "Farmers don't like to see that marketing charge in there, they think they can do it on their own, but they're not seeing the bigger picture."

Several members reported farmers breaking contracts as corporations pay extra high prices to secure their supply. The relationships that OFARM is forming will be crucial when the global supply of these commodities grows and exerts a downward pressure on price—a process already under way in the United Kingdom according to Smith and Marsden (2004), but product needs to move through OFARM now in order to establish those relationships and nurture the overall network.

As a result, communicating the long-term value of OFARM to the farmer members of member organizations has been a constant issue. One of the questions that the organization will have to resolve is how to approach its own promotion. Until now, OFARM has expanded because of the growth of member organizations and the addition of a new organization that incorporated in December 2004. Now, many in OFARM want to see the organization make more connections with existing organizations and organic growers that are not yet part of an organization. OFARM is still making promotion decisions (this conference or that? this publication or that? what message to send?) on a piecemeal basis, rather than as part of an overall strategy.

Most interviewees emphasized that OFARM must grow to build on its initial success. "We need growth and financial stability," one member asserts. "There is a lot we could do, but we need the funding and personnel to do it." Another comments that "it's been a struggle to get the word out, to get other producers to come on board."

Ultimately, the goal is to organize more specific commodity groups within OFARM, which entails recruiting members from dairy, livestock, vegetable, and other sectors. Having proved itself worthwhile, OFARM now has to capture the resources that will enable its leadership role in the organic industry. Toward that end, OFARM has been staying apprised of current debates about exemptions of organic producers from conventional federal and state commodity checkoff programs.

Finally, OFARM faces constraints stemming from the legally circumscribed structure of the organization. To comply with the Capper-Volsted Act, which exempts farmers from antitrust legislation, all member organizations of OFARM have to be composed entirely of working farmers whose products represent at least 50 percent of that organization's sales. This law is based on the assumption that all farmers have the same interests and that farmers' interests are naturally opposed to those of others in the food chain. Yet OFARM discussions frequently center on the challenges of sustaining the entire organic industry along with independent businesses up and down the chain in a time of short supply and corporate encroachment. In this sense, the interests of OFARM are more closely aligned with independent downstream businesses than with corporate farms. Dairy farmers and livestock producers are among the biggest customers of OFARM members because they need grain for feed. OFARM is working on bringing these producers under the umbrella in order to better maintain an industry that supports independent production. But other kinds of businesses in the value stream (processors, bakeries, etc.) are not eligible to participate under the Capper-Volsted requirements.

Discussion

On the most basic level, we conclude that OFARM is an interesting case that represents a promising organizational and business model for creating more spaces for producer autonomy in a more democratic food system. Our preliminary research found that OFARM's strategy of organizing producer groups to stabilize the organic commodity market and maintain a consistent, viable farm-gate price reflects a commodity systems perspective among its participants. OFARM does not seek simply a higher price but rather a more powerful role for producers in shaping a fair and functioning industry that sustains family-scale producers at all points in the chain. As OFARM members report, the boom in organics and now especially organic animal products (dairy, eggs, and meats) has not been an unqualified good for OFARM. While the trust and respect among current participants enables them to pursue distinct collaborative strategies, the unmet demand for organic commodities means that giant food corporations that are getting into organics (like Hood and Dean in dairy) are offering extra-high prices to farmers, reducing the flow of product through these organizations.

The case of OFARM highlights a facet of the conventionalization discussion that has not received as much attention: the participation of field crop growers (and other formerly conventional growers) in shaping the new value chains. As others have commented, close examinations of value chains show the limits of the conventional-alternative distinction in describing how these chains are formed and sustained (see, for example, Hall and Mogyrody 2001; Ilbery and Maye 2005; Whatmore and Thorne 1997). As such, closer attention is being paid to the governance of these networks and the power relations within (Raynolds 2004). Our results suggest that a term like conventionalization obscures the root of the problem: the undemocratic concentration of control within an industry that enriches few while impoverishing or excluding many. Instead of condemning commodified practices in farming as insufficiently alternative, consumers and activists would do well to forge common cause with those seeking to democratize the food system.

Most broadly, these results add new evidence to the foundational idea of materialism: that social structures shape economic processes and outcomes, which means that abstract neoclassic economic assumptions will tend to ignore the contingencies that shape markets. In this case, for instance, the seemingly inexhaustible and rising demand for organic food products (made effective by the growing corporatization of the organic industry) carried both benefits and detriments for organic producer groups. Demand in excess of supply has allowed producer associations to cooperate more fully and establish trusting relationships. It has strengthened the hand of organic producers in relationships with buyers as well. Yet it has also created conditions that threaten to undermine the effort to organize at exactly the time when such efforts are critical to sustaining family-scale farming in the long run.

Our results also suggest that students of agrifood systems are encountering many of the same questions as students of industrial and high-tech development, and finding similar contours and concerns. The emphasis on trust, knowledge sharing, flexibility in organization, and regionalization from the OFARM case are strongly reminiscent of the flourishing literature on industrial agglomerations. One key difference is whereas that literature focuses on innovations in production, OFARM members stress gaining market power and participating in shaping the industry on more democratic terms. In that way, the case of OFARM is more closely linked to the commodity system literature, which has yet to be synthesized with the industrial district literature. Further research on OFARM and similar cases would fruitfully draw on both lines of research.

Finally, our results point to two directions for future research on OFARM and organic field crops in general. First, to use William Friedland's (2004) terminology, we aim to turn this "commodity analysis" into a "commodity systems analysis" by seeking more complete information about the entire value chain of organic field crops, particularly the link to organic animal agriculture via the buyers that OFARM members deal with, and from there, to the moment of retailing (Smith and Marsden 2004). Second, more horizontally, to address the question of agrifood alternatives and rural development, we seek to compare an organizational model like OFARM with others, current and historically, that similarly seek (or sought) to sustain family-scale producers in the face of corporate concentration. Historically, for example, the efforts to organize small-scale fruit growers in the Northwest (which inspired the original Capper-Volsted legislation) would likely yield some important lessons for contemporary efforts. A current case of a similar effort is produce auctions (Swanson 2005) that help small-scale producers sell to medium-scale buyers without the legal, financial, and organizational encumbrances of a cooperative. Regardless of specific research strategies, the story of OFARM encourages us to remember, in the language of Howard Becker (1998, 46), that things are just "people acting together," and it's people that count.

Note

The research for this chapter was supported by a grant from the Leopold Center for Sustainable Agriculture. Iowa State University, Ames. An earlier version of this chapter was presented on August 8, 2005, at the RC–40/Sociology of Agriculture miniconference titled An Agriculture without Subsidies? Visioning a Market Driven Agriculture, in conjunction with the sixty-eighth annual meeting of the Rural Sociological Society, Tampa, Florida. The chapter is largely based on portions of Amy Guptill, Richard A. Levins, Laurance R. Waldoch, and Rick Welsh, "Forming Agricultural Bargaining Units for a Sustainable and Equitable Agriculture: The Case of the Organic Farmers Association for Relationship Marketing," report to the Leopold Center for Sustainable Agriculture, October 2005.

References

Becker, Howard S. 1998. *Tricks of the Trade: How to Think about Your Research While You're Doing It.* Chicago: University of Chicago Press.

Blank, Steven C. 1998. *The End of Agriculture in the American Portfolio.* Westport, CT: Quorum Books.

Braun, Boris, Wolf Gaebe, Reinhard Grotz, Yoshiyuki Okamoto, and Kenji Yamamoto. 2002. Regional Networking of Small and Medium-Sized Enterprises in Japan and Germany: Evidence from a Comparative Study. *Environment and Planning A* 34:81–99.

Britton, John N. H. 2003. Network Structure of an Industrial Cluster: Electronics in Toronto. *Environment and Planning A* 35:983–1006.

Buck, Daniel, Christina Getz, and Julie Guthman. 1997. From Farm to Table: The Organic Vegetable Commodity Chain of Northern California. *Sociologia Ruralis* 37:3–20.

Coombes Brad, and Hugh Campbell. 1998. Dependent Production of Alternative Modes of Agriculture: Organic Farming in New Zealand. *Sociologia Ruralis* 38:127–145.

Cumbers, Andy, Darny MacKinnon, and Karen Chapman. 2003. Innovation, Collaboration, and Learning in Regional Clusters: A Study of SMEs in the Aberdeen Oil Complex. *Environment and Planning A* 35:1689–1706.

Friedland, William H. 2004. Agrifood Globalization and Commodity Systems. *International Journal of Sociology of Agriculture and Food* 12:5–16.

Green, Gary P. 1994. Is Small Beautiful? Small Business Development in Rural Areas. *Journal of the Community Development Society* 25:155–171.

Grow, Shelly, Amy Guptill, Thomas A. Lyson, and Rick Welsh. 2003. The Effect of Laws That Foster Agricultural Bargaining: The Case of Apple Growers in Michigan and New York State. Available at ⟨http://www.winrock.org/GENERAL/Publications/AgBargfinal.pdf⟩.

Guptill, Amy. 1998. Beyond California: The Growth of Small Wineries in the United States. Master's thesis, Cornell University.

Guthman, Julie. 2004. The Trouble with "Organic Lite" in California: A Rejoinder to the "Conventionalisation" Debate. *Sociologia Ruralis* 44:301–316.

Hall, Alan, and Veronika Mogyrody. 2001. Organic Farmers in Ontario: An Examination of the Conventionalization Argument. *Sociologia Ruralis* 41:399–422.

Heffernan, William D. 2000. Concentration of Ownership and Control in Agriculture. In *Hungry for Profit: The Agribusiness Threat to Farmers, Food, and the Environment*, ed. Fred Magdoff, John Bellamy Foster, and Frederick H. Buttel, 61–75. New York: Monthly Review Press.

Hendrickson, Mary, William D. Heffernan, Philip H. Howard, and Judith B. Heffernan. 2001. *Consolidation in Food Retailing and Dairy: Implications for Farmers and Consumers in a Global Food System*. Report to the National Farmers Union, June 1.

Hinrichs, C. Clare, and Rick Welsh. 2003. The Effects of the Industrialization of U.S. Livestock Agriculture on Promoting Sustainable Production Practices. *Agriculture and Human Values* 20:125–141.

Ilbery, Brian, and Damian Maye. 2005. Alternative (Shorter) Food Supply Chains and Specialist Livestock Products in the Scottish-English Border. *Environment and Planning A* 37:823–844.

Kenney, Martin, and Donald Patton. 2005. Entrepreneurial Geographies: Support Networks in Three High-Tech Industries. *Economic Geography* 81:201–228.

Levins, Richard A. 2001. *An Essay on Farm Income*. Staff paper P01–01, Department of Applied Economics, College of Agricultural, Food, and Environmental Sciences, University of Minnesota.

Lobao, Linda, and Katherine Meyer. 2001. The Great Agricultural Transition: Crisis, Change, and Social Consequences of Twentieth Century US Farming. *Annual Review of Sociology* 27:103–124.

Lyson, Thomas A. 2004. *Civic Agriculture: Reconnecting Farm, Food, and Community*. Medford, MA: Tufts University Press.

Lyson, Thomas A., and Charles C. Geisler. 1992. Toward a Second Agricultural Divide: The Restructuring of American Agriculture. *Sociologia Ruralis* 32:248–263.

Lyson, Thomas A., and Amy Guptill. 2004. Two Types of Farming in the U.S. Today: Commodity Agriculture and Civic Agriculture. *Rural Sociology* 69:370–385.

Lyson, Thomas A., Amy Guptill, and Gilbert W. Gillespie Jr. 2000. Community Engagement and Dairy Farm Performance: A Study of Farm Operators in Upstate New York. *Research in Rural Sociology and Development* 8:309–323.

Lyson, Thomas A., Robert J. Torres, and Rick Welsh. 2001. Scale of Agricultural Production, Civic Engagement, and Community Welfare. *Social Forces* 80:311–327.

Perrow, Charles. 1993. Small Firm Networks. In *Explorations in Economic Sociology*, ed. Richard Swedberg, 377–402. New York: Russell Sage Foundation.

Pietrykowski, Bruce. 1999. Beyond the Fordist/Post-Fordist Dichotomy: Working through *The Second Industrial Divide*. *Review of Social Economy* 57:177–198.

Piore, Michael J., and Charles F. Sabel. 1984. *The Second Industrial Divide: Possibilities for Prosperity*. New York: Basic Books.

Raynolds, Laura T. 2004. The Globalization of Organic Agro-Food Networks. *World Development* 32:725–743.

Sabel, Charles F. 1992. Studied Trust: Building New Forms of Cooperation in a Volatile Economy. In *Industrial Districts and Local Economic Regeneration*, ed. Frank Pyke and Werner Sengenberger, 215–250. Geneva: International Institute for Labour Studies.

Smith, Everard, and Terry Marsden. 2004. Exploring the "Limits to Growth" in UK Organics: Beyond the Statistical Image. *Journal of Rural Studies* 20:345–357.

Staber, Udo. 2001. The Structure of Networks in Industrial Districts. *International Journal of Urban and Regional Research* 25 (3): 537–552.

Strauss, Anselm, and Juliet Corbin. 1998. *Basics of Qualitative Research: Techniques and Procedures for Developing Grounded Theory*. 2nd ed. Thousand Oaks, CA: Sage Publications.

Swanson, Mark A. 2005. Produce Auctions and Local Food Systems: Reinventing Distribution Networks. Paper presented at the joint annual meeting of the Agriculture, Food, and Human Values Society and the Association for the Study of Food and Society, Portland, Oregon, June 9–12.

Tolbert, Charles M., Thomas A. Lyson, and Michael D. Irwin. 1998. Local Capitalism, Civic Engagement, and Socioeconomic Well-Being. *Social Forces* 77:401–428.

van der Ploeg, Jan Douwe. 2000. Revitalizing Agriculture: Farming Economically as Starting Ground for Rural Development. *Sociologia Ruralis* 40:497–511.

Volkin, David. 1995. *Understanding Capper-Volsted*. Cooperative Information report 35, U.S. Department of Agriculture Rural Business and Cooperative Development Service.

Welsh, Rick. 1996. *The Industrial Reorganization of U.S. Agriculture: An Overview and Background Report*. Policy Studies report no. 6. Greenbelt, MD: Henry A. Wallace Institute for Alternative Agriculture.

Whatmore, Sarah, and Lorraine Thorne. 1997. Nourishing Networks: Alternative Geographies of Food. In *Globalising Food: Agrarian Questions and Global Restructuring*, ed. David Goodman and Michael Watts, 287–304. New York: Routledge.

5

Contractual Integration in Agriculture: Is There a Bright Side for Agriculture of the Middle?

Mary Hendrickson, William Heffernan, David Lind, and Elizabeth Barham

[handwritten margin note: Key to success is protecting marketing access]

The key to success for the agriculture of the middle is in protecting market access for farmers. Today, farmers in the middle are operating in an agrifood system that is continuing to undergo increased vertical integration (Heffernan et al. 1999; Hendrickson et al. 2001; Welsh 1998) as well as decreasing competition at all stages of the food system. Contracts are part of an agricultural system no longer characterized by competition—a system in which no single buyer or seller buys or sells enough to influence the market price. This change has threatened family farms—those in which the family provides the majority of the management, labor, and capital for the farm—because they are dependent on a competitive marketing structure. In a globalizing food system, farmers will find that they need to contract with an already-established global food system cluster (Heffernan et al. 1999), or develop their own short production and processing chains with new allies in order to have a market.

[handwritten margin note: Contracts can be good or bad]

In this chapter, we examine the conditions leading to contracting in agriculture and the changing power relationships that can evolve in contractual arrangements when contracts become a prominent feature of agricultural systems. We make no arguments about whether contracts are good or bad for agriculture. Our focus is on the changing nature of the power relationships involved in agricultural contracting. For us, the key to determining whether contracts can enhance the socioeconomic well-being of farmers of the middle is the position of power from which these farmers operate. If two parties negotiate a contract from equal positions of power, the contract can reduce uncertainty and perhaps transaction costs for both parties. For farmers in the middle, accepting or rejecting contracting will depend on their positions of power to negotiate with other firms in a value chain or food system cluster. From this perspective, an understanding of the historical nature of contracts and their role in agriculture provides a useful point of departure for policy recommendations aimed at assisting agriculture of the middle.

Defining Contracts in Agriculture

Contracts of varying types have a long history in agriculture. The major change in contracting, however, focuses on contracts as they connect the production stage to the supply and processing stages of the food system—or what we once called the market. Traditionally, a distinction has been drawn between marketing and production contracts. Marketing contracts are sales agreements between a producer and buyer that determine the price, quality, quantity, and time of delivery of a product before it is produced. They are compatible with family farms *as long as there is still competition in the market*, because then farmers have a range of choices with whom to contract. An important aspect of marketing contracts is that the farmer owns the product, makes most of the major managerial decisions regarding production, and seeks to sell their product for a price that will cover the cost of labor and investments. Marketing contracts are only a slight move away from the "spot market" in which farmers take the products produced to the market where they can get the best price with no preexisting agreement as to who will buy the product.

The production contract, characterized by integrating firms and growers, is the form of contracting that dominates the poultry and hog industries today. The poultry industry has now become the prototypical model of industrialized agriculture and is often referred to as a model of the structure that may come to characterize much of U.S. farming in the future (Perry et al. 1999; Hendrickson et al. 2001). Production contracts are central to the industry's organization and development. Nevertheless, the production contract is really not an arrangement to exchange ownership of an agricultural good because the product—for instance, broilers—is never sold. The integrating firm owns the breeding flocks, the unhatched eggs, and the chicken its entire life, and may not exchange it in a market until it is sold to a wholesaler and/or retailer as fresh poultry or a TV dinner. Because the grower never owns the product, there is no "price" negotiation for the broiler as in marketing contracts. The price negotiation is really about the price of labor—labor that happens to bring assets to the relationship in the form of buildings, equipment, and land.

The History and Economic Theory of Contracts

From a sociological perspective, the emergence of contracts marks the transition from traditional, status-based societies to modern democracies and markets. As feudal systems of rule declined, contracts helped to bring order and legitimacy to the interconnected relationships between the individual, the state, and the economy by providing rational practices for establishing agreements and obligations between political equals. Contracts were also important in the process of rejecting the legitimacy of inherited privilege. The contract shifted political life negotiated through the norms and constraints of interpersonal relationships to the public realm of the state. Doing so promoted "the creation of a legal personality, individual and inalienable rights, public accountability, and collective bargaining" (Watts 1994a, 25). The notion of individuals in a free society with minimal overrule and maximized choice is at the heart of the concept of contract, and is central to modern democracies and markets (Wallerstein 2000). The attraction of a contract is that ideally, it involves making a voluntary agreement between political equals that constitutes the legal foundations of claims and obligations between individuals in civil society as well as between civil society and the state. This is exactly the assumption most economists make when discussing contracting.

While contract is an important feature of our political history, the contract is also "the characteristic legal institution of an exchange economy which is strongly market oriented and based on the use of money" (Fox 1974, 153). In the United States, nineteenth-century economic expansion led to the full development of contract law. During this period's dynamic growth, the fundamental organizing principle was the protection of a "freedom of contract" (Teeven 1990) that sought to minimize state oversight in contractual relations. Following the economic crisis of the early twentieth century, however, the perceived failure of laissez-faire practices led to antitrust, labor, and insurance laws that limited contractual freedoms. The government became much more active in managing business cycles as a result of new fiscal and monetary policies, but it also began to use its powers to affect wage relations and the rights of workers. This period was marked by a tense but "firm balance of power," or "social contract," between the three social institutions—the state, corporate capital, and organized labor (Harvey 1990, 133–134).

Today, the contract is leading to a social and political transformation in the lives of rural producers. Michael Watts (1994b, 255) has suggested that the proliferation of contracting in agriculture around the world speaks to the genesis of a new "social economy" of uneven relations between rural producers and agribusiness firms. Philip McMichael (1999) has used the example of contracting in agriculture as one illustration of how contemporary agribusiness firms increasingly depend on nonwage forms of labor. He argues that the wage form governs value production less and less, and that nonwage forms of labor are becoming at once more significant and increasingly tenuous. McMichael suggests that the proliferation of nonwage forms of labor today may define a historical period in which the wage labor contract is eroding in the face of an attempt to elaborate a politicized form of market rule on a global scale.

In summary, contracts can best be understood in the context of the political and economic structures within which they are practiced. The globalization of agricultural markets, with its evolving contract relations, heralds an important new context within which social, political, and economic relationships are developing. While contracting is not new to agriculture, researchers argue that the contemporary context of production contracts suggests a more general shift in the political economy of agriculture that hinges on "politically mediated relations between the state, firms and labor within a global economy" (McMichael 1999, 19). In this context, the contract increasingly acts as a "policing mechanism" (Wolf et al. 2001) employed by powerful agribusiness corporations to enforce the rules and obligations of private law, rather than as a democratic process of mutual determination between political equals.

Economic Structure and Contracts

Economic approaches to contract tend to derive from economic agency theory, and view the contract as a relationship between a principal (the contractor, typically the agribusiness firm) and an agent (typically the grower). It assumes that the party who performs a function will receive a reward from the principal, the party that instigated the contractual relationship. One of the problems that principals negotiating a contract face is the difficulty of observing and verifying the efforts of the party performing the activity. This creates a situation in which information is "asymmetrical" (see Wolf et al. 2001).

This asymmetrical information creates the possibility that the person performing the function will behave opportunistically either by "shirking" effort, or underinvesting in necessary inputs and assets. Under these circumstances of uncertainty, the main concern of the agribusiness firm is to organize and control grower behavior in such a way as to reduce uncertainty and maximize net returns (Johnson et al. 1996). As Steven Wolf and colleagues (2001) describe it, the contracting firm reduces its uncertainty through a set of interlocking enforcement or policing mechanisms through which the firm is able to better monitor and discipline growers by subjecting them to efficiency and production criteria.[1]

A great deal of attention has been given to explaining the strategic rationale behind contracts and the role the principal (the integrating firm) plays in the production process. Little attention, however has been focused on how one party to the contract acquires the dominant role, and thus the right to command or police the behavior of the other. Debates about contracts in agriculture often differentiate between contracts based on the dynamics of ownership as if different ownership arrangements are a legitimate basis for different management rights. For example, marketing contracts are distinguished from production contracts in that farmers own the product (chicken, grain, etc.) in the former, but do not own the product in the latter. Consequently, the extensive management control in the production contract is usually justified on the grounds that the grower is handling the integrating firm's property and therefore must follow the directives of the integrating firm. The broiler sector is an interesting case because growers are responsible for about one-half of the capital invested in the entire broiler production and processing stages (Heffernan 1998). Of course, all of their capital is invested in the growing of birds, which gives them by far the greatest capital investment in the production stage. Yet the growers take orders from the integrating firm.

Economist David Ellerman (1992) argues that contractual roles—that is, who is the principal, and who is the person receiving the orders—are not determined by property rights or some other inherent right of the parties to an agreement. Contractual roles are the outcome of a "hiring contest," and while corporate capital tends to do the hiring, this is a consequence of its market power in the hiring contest rather than a property right. From this economic perspective, market power is formalized through the contractual agreement, and the right to command, police, or manage is seen to derive not from ownership but from winning the contest that determines the "legal" identities of the contracting parties.

How does one acquire the positive control right over another person's behavior—the right to tell them what to do? Hire them.

Thus, the right of command is a managerial right. In the case of broiler production, the focus of the contract is on managing labor. Growers are paid on a piece rate of four to five cents a pound plus various incentive measures, and for this the integrating firm gets the right to dictate the criteria of the building and equipment the grower must provide. The contract is usually silent on how much is allocated to the assets that the grower provides. Use of the assets is implicit in the labor contract that has been constructed, and the integrating firm uses the assets of the grower with other factors to produce goods. Property rights do not explain why the integrating firm has the right to control decisions about the grower's assets.

[handwritten margin note: "Control over behaviour by 'hiring']

In markets with high levels of concentration, like most of the markets that farmers of the middle depend on (Hendrickson and Heffernan 2002), market power along with its consequences for bargaining power determine who is the principal and who has the right of control. Market power makes it possible for the integrating firm to win the hiring contest and take on the legal role of principal. The prize of this contractual battle is the legal or contractual right to manage or control the grower as well as the legal or contractual right to claim the finished product (and profit).

Union Parish, Louisiana: A Thirty-Year Case Study in Contracts

In marked contrast to the informal contracts typically used fifty years ago, the production contracts that are used today in the poultry sector are formal, and spell out in great detail the rights and responsibilities of the parties involved. While contracts vary from company to company and commodity to commodity, a typical production contract outlines the roles and responsibilities of each party, and provides for a variety of incentives and penalties based on management performance and quality standards. In the broiler industry, the grower is responsible for the labor and the fixed capital investments in land, buildings, and equipment needed to care for the birds. The typical grower is also responsible for utilities, manure and dead chicken disposal, chicken house cleaning, and other operating expenses, such as repairs and maintenance. The live chicken is at no point the property of the grower. The integrating firm owns the birds unless they die on the grower's property, at which point they become the grower's responsibility to dispose of the carcasses.

The integrating firm provides the birds, feed, and usually some of the health products. Thus, the integrating firm determines the genetics that will be used, the specific feed ingredients, and the timing and frequency of when the grower receives the birds. It sends a catching crew to the building when the birds are to be processed. It also outlines feeding schedules, and supplies management services (field persons) to help for most unplanned events, such as when the water system fails or temperatures are extremely high. Typically, the integrating firm specifies the capacity and construction of the chicken houses as well as the technologies used in production, such as automated water and feeding systems.

Poultry is different than many other commodities for a variety of reasons, but an important difference lies in poultry rations requiring relatively high protein and a delicate balance of amino acids relative to ruminants. This made broiler production more dependent on commercial feed firms, which became the major consolidators in the broiler industry. As the first integrating firms, they manufactured the feed, sent it by truck to growers, picked up the matured broilers, and transported them to their slaughtering facilities located near the feed facility. In the early years, there were hundreds of integrating firms that owned both feed and slaughter facilities. They would ordinarily travel twenty-five to thirty miles out in a circle from the processing plant to the growers' buildings (Heffernan 1984). The geographic layout is much the same today except that the number of integrating firms and processing facilities are greatly reduced. There are currently only about fifty integrating firms, and the largest four have 55 percent of the market share. These firms have about 250 sets of processing facilities across the country producing broilers.

Few growers live in an area where two circles of competing integrating firms overlap. As a result, most growers live in places where they have access to only one integrating firm.

A longitudinal study begun in 1969 provides an opportunity to examine the changes that occurred in the structure of the broiler sector in one geographic region, the changing relationship between the integrating firm and the grower, and the consequences of these changes during the last half of the twentieth century. As part of a larger study focusing on structural change in agriculture, Union Parish, Louisiana, was selected as the site for the first phase of this study because it had a sample of poultry growers, family farms, and larger than family farms in the same cultural, social, and geographic region. It also had more contract broiler producers than any other county in the state—a position it still holds today. In 1969, Union Parish lagged behind most Louisiana parishes in the value of sales of farm-produced commodities, but with a doubling of the number of poultry growers, it was among the top counties in terms of gross farm sales in the state by 1999 (USDA 2002). Yet Union Parish was identified as a persistent poverty county in 1969, and remained so thirty years later. Clearly, integrated broiler production—a system that Louisiana borrowed from areas on the East Coast—has not been a rural economic development success story.

In 1969, all growers (N = 53) and family farmers (N = 24 in Union and N = 25 in Parish B) in the parish were interviewed. At that time, there were four processors operating in the parish—two that were locally owned and operated, and two that were owned by integrating firms with headquarters outside the parish. Growers at the time had potentially four choices of processors, and the perception was that there was competition in the broiler business.

During structured interviews in 1969, growers shared information on the changes in broiler production during the previous decade. Beginning in the 1950s, following the scientific conquering of communicable diseases in poultry production, and the beginning of vertical integration in the industry, the proprietors of the local feed supply firms entered into informal agreements with some local farmers. The proprietors of the feed supply firms agreed to provide the feed and later the birds if the farmer would construct a building using the latest technology, and supply the land and labor. Even by 1969 standards, some of the original

Table 5.1
Grower Ratings of Their Integrator*

	1969	1981	1999
Poor	—	1%	18%
Not so good	10%	8%	35%
Good	63%	72%	43%
Very good	24%	19%	4%

* As a company to do business with

buildings looked primitive, but they represented a movement toward environmentally controlled buildings that would eventually lead to large capital requirements. Entering into the agreements as equals, the feed suppliers and farmers shared in the expenses and profits.

The early agreements were often verbal, and between local people who had kinship and friendship ties. As some firms became larger and located farther away from the communities of the growers, the informal agreements were replaced with formal codified contracts. These contracts spelled out in more detail the rights and privileges of all parties in the economic contract.

In 1969, the growers were quite satisfied with their broiler operations (see table 5.1) and the economic benefit it brought to their family, and they had favorable relationships with and trusted their integrating firm. Interestingly, they were already beginning to realize a difference in the decision-making opportunities between those raising only broilers and those producing commodities other than broilers. Those involved only in broiler production were beginning to describe themselves as growers rather than farmers, already recognizing their new position. Most of those who added the broiler enterprise to other agricultural enterprises still perceived themselves as farmers even though they were broiler growers.

In 1981, contract broiler growers (N = 57) were again interviewed. At that time, only two processing firms—one national and one international—remained available to Union Parish growers, since the two local processors had ceased operating as integrating firms in the 1970s. This phase of the research found that the quality of life for most grower families had improved in the twelve-year period. Seventy-seven percent of the growers indicated that their financial condition had improved over the previous ten years. At the same time, with only two integrating firms left in the parish, growers were becoming increasingly concerned with the power relationship between themselves and the firms. In 1982, shortly·after the second study was completed, the two remaining firms merged and for the next fifteen years there was effectively only one processor available to broiler growers in Uniòn Parish (Heffernan 1984).

The number of firms available to growers in Union Parish has been important in setting the context within which the power relations between growers and integrating firms evolved. In the early years, the feed suppliers and the growers were each operating in a somewhat competitive environment. In this environment, where each party had the opportunity to accept or reject the terms of the contract because there were other suppliers or buyers available, the parties entered the contract from a relatively equal basis of power. (Even then an informal norm was evolving between integrating firms to not raid their competitors' growers.) When the two local firms, the ones in which some growers placed more trust, ceased operation and only two integrating firms remained, the growers began to be concerned about their increasing powerlessness. Even though they were still doing quite well financially, they knew that they were no longer political equals. Once the two firms were merged, the growers had access to only one processing facility within twenty-five to thirty miles of their location.

The third phase of the study was carried out in 1999. The fieldwork began with a group of 12 growers identified by the local Contract Poultry Growers Association and expanded to 118 broiler growers identified in the parish. By 1999, a broiler house that met all the company's specifications for a Class A facility—for example, a climatically controlled chicken house with insulated curtains—cost approximately $125,000— an increase of about $45,000 since 1981. While broiler operations could get by on two to three houses in 1981, today the average grower in Union Parish operates four houses. This number is in accordance with recommendations from both growers and creditors in northern Louisiana who advise prospective growers that a viable operation requires a minimum of four buildings to create the cash flow necessary to pay off the investment. Based on these figures, a prospective grower in the parish would need to build at least four chicken houses to company specifications, amounting to a $500,000 investment—mostly borrowed—before being able to secure a contract with the one remaining integrating firm.

Lessons from Louisiana for Agriculture of the Middle

[handwritten: Imbalances of power]

Debt and Power

[handwritten: ① debt causes a huge imbalance of power]

Debt is probably the most important factor leading to the inequality in the power relationship between growers and the firm when the grower has access to only one processor. Most growers begin the contractual relationship owning their farms, but they have few other economic assets. They must borrow most of the $500,000 required to construct and equip the buildings, and the only assets they have for collateral is their farm, which typically includes their home. This makes the grower totally dependent on the integrating firm for chickens, feed, a fair payment for their work, and use of their assets while growing the chickens. On signing the contract, the grower's livelihood becomes tied to fixed capital investments and the economic relationship with the firm. Because the integrating firm can hire other growers in the area or near their processing plants somewhere else in the country, they can easily break the relationship with any one grower. The firm's survival does not depend on any single grower, but the grower's survival and life's assets does depend on a single integrating firm. An integrating firm's survival does, of course, depend on a collectivity of growers, leaving open opportunities for collective action on the part of growers.

In 1999, 79 percent of the growers were in debt and 21 percent were debt free. A higher proportion of growers are out of debt now than ever before, but probably not because of economic conditions. A number of other factors also account for this improvement in the debt to nondebt ratio. First, those growers who are out of debt have been in the business an average of ten years longer than those with debt. On average, those out of debt have been growing chickens for twenty-four years, and in 1981 few growers had been producing that long. Second, those who are out of debt on average incurred a much smaller debt when they began ($92,000 as compared to $240,000).

The initial capital investment is not the only capital investment that growers must make. The growers must also provide for maintenance

Table 5.2
Union Parish Growers and Debt

	1981 (N = 57)	1999 (N = 112)
Growers with debt	54	75
Growers without debt	3	20

and repair. In addition, new technologies continue to be developed that can improve the efficiency and profit for the integrating firm. Growers in Union Parish describe a cycle of technological changes in which every few years the firm revises its specifications on houses and equipment, and pushes growers to readjust to the new requirements. In 1999, growers were being encouraged to update their houses. Since 1995, 30 percent of the growers had updated 101 houses at an average cost of $24,000 per house—slightly less than the average net annual income from four broiler buildings ($27,500). Many growers complain that just as they are about to get their debt paid off, the firm requires them to make major changes in their buildings that require the grower to once again go deeper into debt.

As growers discuss their experiences with changing equipment specifications, it becomes apparent that despite occasional monetary incentives provided by the firm, they view an insufficient return on their new investments as a serious problem. As one grower explains:

The cost of repairs and equipment has run up but income hasn't. That is what puts so much strain on people. The company is continually changing requirements. It is a revolving door. You can't keep up. Two years ago I put a new water system in all five of my houses, because the company said to do it. About a year later, the requirements changed. That watering system is buried in the ground and I'm still paying for it.

Even when growers recognize the advantages offered by innovations, they have a sense that they are the guinea pigs in a relationship that requires, but does not always compensate, their additional expense.[2] Some growers conclude that experiments in new technology serve to keep the grower locked into the contract relationship. One grower asserts that "we had debt and you couldn't get out of it. That is the way [the firm] wants you. If they can keep you in debt they can control you."

up farmers in debt

Liquidity of Assets and Power

The second issue leading to an imbalance in the power relationship between the growers and the integrating firm relates to the type of assets each supplies. Most of the growers' assets are what are called fixed capital, while the integrating firm invests in movable capital. Fixed capital refers both to the spatial and relational rigidity of investments in land, buildings, and specialized technology. Fixed capital is located in a place, and cannot be easily picked up and moved. It also refers to those qualities of technology, especially specialization, that create specific and dependent relationships. As technology has become more specialized in the farming sector, its purpose and use has become increasingly specific. Oliver E. Williamson (1991) describes this as "asset specificity," a situation similar to the idea of sunk costs that refers to the degree to which an asset can be redeployed to alternative uses or users without sacrificing the assets productive value. For example, in the broiler industry, chicken houses are constructed for a specific purpose—growing chickens—and have little use outside that purpose.

Because the integrating firm has the option of not entering into a relationship with a new owner of the broiler facilities, the firm can effectively block a meaningful sale of the property. Once an asset-specific investment has been made, there is only one way to get a return on the investment. Although the integrating firm has some capital in fixed assets at its processing site, in the production stage its capital is movable. Movable capital refers to those things that move through the food system. In the case of broilers it includes feed, birds, meat, and such branded products as TV dinners. These products can be turned into cash on short notice.

Length of Contract versus Length of Mortgage

Another issue that creates a major risk for the grower that in turn leads to more inequality between the grower and the integrating firm is the time period required to repay the debt relative to the time period covered by the contract. Going into debt is an investment in a business, but it is also an investment in a relationship that is formally defined by the production contract. The investment in the relationship is formed by the strategic and political circumstances (that is, the structure) in which contracting is conducted that can change over time. For instance, having made a $500,000 investment, the grower expects to have a long relationship with the integrating firm, but current poultry contracts do not assure this. In fact, while early poultry production contracts were written for several years, today such contracts do not exist. In 1969, some growers were able to obtain contracts defining the number of birds per year and the price per pound for up to seven years. They were also able to get loans for up to seven years that could be repaid from the profits generated from the contracts.[3] The contract reduced the risk for both the firm and the grower. Over time and with the loss of three of the four firms, the period covered dropped from seven years to the length of time corresponding to that of the broiler's life (about six to seven weeks) or what is called "batch to batch."

The firm greatly reduced its risk because it did not need to guarantee a minimum price to growers for several years into the future and could reduce the payments to the growers for whatever the reason. In addition, the integrating firm could reduce its uncertainty regarding whether or not it would have enough producers over the long term to meet market needs. To be sure, the integrating firm faces some limits because it must be assured that it does not bankrupt all of its growers. On the other hand, the risk to the grower has increased substantially. The grower may have reduced the risk of not finding a market, but the risk of being locked into debt with no assurance of receiving a payment large enough to cover all expenses and a reasonable payment for labor has greatly increased. In fact, the grower has picked up much of the risk that the firm was able to escape. Thus, the grower's risk is not only the large debt backed by a mortgage on the farm but the increased potential of economic failure because the grower is totally dependent on one firm for their economic future. The firm is in a position to set the price that the grower receives, and the grower is put in a precarious financial situation.

Returning to Ellerman (1992), it is important to remember that the changes that took place in Union Parish between the integrating firm and the growers were made by changing "the contract." There was a time when the grower and the integrating firm were political equals— when the grower was a *potential* grower. When the integrating firm is seeking to recruit new growers and when a person with land—or one willing to purchase land—is interested in broiler production first come

together, they are reasonably equal.[4] This is when the contract is constructed. The potential grower and the integrating firm remain equals
until the grower borrows the capital investment required and constructs
the building.

Trust and Control

eroded trust

The overall sense of powerlessness that growers feel in relation to contracts with an integrating firm is compounded by an eroded sense of trust
in the company's motives. To maximize their income, growers must
work with the materials they are given by the firm to achieve high levels
of efficiency in feed conversion. When growers treat two or more sets of
birds the same way and the performance is quite different, the growers'
sense of control over farm-level decisions is undermined and many conclude that something is wrong with the birds they were provided.

In 1999, a number of questions were asked that focused on the
growers' trust in the firm's management of various crucial inputs and
measurement procedures that directly affect settlement payments. The
responses suggest that growers question the accuracy of the integrating
firm's settlement reports and the quality of inputs that the integrating
firm supplies. In fact, over two-thirds of the growers occasionally or
often questioned integrating firm reports about the quality and quantity
of birds and feed, and more than three-fifths questioned the number of
birds provided and the condemnation rate. The quality of the inputs
offered by the integrating firm was of particular concern to Union Parish
growers, because the input quality comes to symbolize both their lack of
control over their operation and the difficulties they have placing their
trust in the integrating firm. "You have basically put all your trust in
what the company hands you. If they don't have good [genetics], we
will have sorry chicks and no matter what you do there isn't much you
can do about it." Concerns with the quality of their chickens and feed
along with the accuracy of integrating firm measurements are compounded by the fact that growers feel powerless to negotiate disputes.
As one grower observed, "You can't prove anything. You have to take
their word for it. You have to show them where the feed went. You can't
put scales on the farm. Grin and bear it, [because] it doesn't do you any
good to raise questions. You couldn't do anything with the company if
you tried."

Table 5.3
Grower Trust in Integrator Practices*

Quality of birds	86%
Quality of feed	72%
Weight of feed	68%
Weight of birds	67%
Number of birds	63%
Condemnation rate	62%

* As measured by percent of growers who occasionally or often question integrator reports of various factors relevant to production and settlement wages

In Union Parish, the lack of market options for growers has led to a situation where many growers don't even read their contracts because they have no sense they can negotiate. For these growers, signing the contract becomes little more than a symbolic act of submission: "They change the contract every year. I've even had the serviceman rush me the contract to sign as the chicks were pulling up to the houses. Half the time I don't even read them." As another grower put it, "I have been signing them without reading them because they have to be signed."

Control and Power

Over time, the contract has changed farming and the farmer. Through contracting, farmers are transformed into a sufficiently disciplined labor force to ensure the profitability and survival of the integrating firm (Watts 1994a, 255; Wolf et al. 2001). In the early 1980s, firms in the broiler industry were primarily concerned with organizational survival.[5] They needed to recruit more growers as well as maintain a good relationship with the growers they already had (Heffernan 1998). At that time, Heffernan (1984, 257–258) predicted that as fewer firms gained control of the broiler sector and their survival was more assured, managers would be able to place more emphasis on profits. One way for the firms to increase profits has been to reconfigure the contract into a distinctive labor system in which the growers' labor is increasingly marginalized, devalued, and subordinated to the prerogatives of a politicized form of market rule.

Under the production contract, labor is subordinated to the firm's profit margin in two ways. First, as we have demonstrated, labor (the grower) is made responsible for all fixed capital investments in land, buildings, and equipment. The grower is therefore disciplined by the debt incurred on entering broiler production, and the firm is released from the costs of fixed capital investments in the production stage of the broiler. Second, the growers' settlement payments are figured on a piece-rate basis and subject to a "tournament"—a company-designed competition among a group of growers that determines their penalties and incentives based on efficiencies in the conversion of the firm's feed into chicken meat. Significantly, such efficiencies are more related to the genetic quality of the birds and the quality of the feed than to the value of the labor. In fact, tournament formulas don't even consider different quantities of labor supplied. The integrating firm has clearly stripped power from the growers by burdening them with debt, providing only short-term contracts with no long-term obligation, and controlling all measures of both quantity and quality of the raw product. Obviously, in this particular context, the power belongs to the integrating firm and not the grower.

Policy, Power, and Contracts

how can farmers maintain sure to many

As horizontal and vertical integration have continued, and consequently the number of input and processing firms to which farmers have access has decreased—in some cases to one—farmers producing without contracts tend to get locked into a food system cluster in much the same way as a grower with a production contract. When a farmer must buy inputs—seed, chemicals, and fertilizers—from the dominant set of firms in a food system cluster and sell products back to the same set of firms because of firm and geographic concentration, the differences between market and production contracts are diminished (Heffernan et al. 1999). Given the nature of contracting and its relationship to the changing structure of agriculture, what options exist for farmers of the middle? How will farmers of the middle retain their autonomy and power, either in producing undifferentiated commodities for globally integrated markets or in joining new supply chains for an emerging food system?

First, there are still competitive—or spot—markets in the food system, found where farmers directly market to their own consumers, such as with pastured poultry production, farmers' markets, or CSA farms.

①

These farmers, operating individually or collectively, are still competing with the vertically integrated food system, but the competition is established on quality attributes and personal relationships rather than price (see chapter 7, this volume).

A second possibility, following Wolf and colleagues (2001), is that two *principals*, rather than one, are necessary if contracts are to benefit farmers of the middle. The key here is for farmers of the middle to act collectively in creating short supply chains (Marsden et al. 2000), where the benefits of production and marketing are shared equitably through the chain. Extrapolating from Ellerman (1992), it is essential for farmers of the middle to remember that there is no inherent reason why farmers—individually and especially collectively—can't become the principal and receive the rights of management. In the real world, capital hires labor most of the time, but when we perceive agriculture as part of a supply chain, farmers are providing capital in the system.

More important, farmers can acquire the right to command by becoming management. When a farmer or a group of farmers takes the initiative to approach a processor and develop an arrangement for processing in which the farmer or group can maintain an identity-preserved product that is processed in specific ways, the farmer or group is management and can maintain some control. This is done by use of a contract. Usually, the management is shared, but control is awarded at this time and property rights have nothing to do with it. If the farmers are skilled negotiators like the broiler integrating firms, they get the use of the processor's assets along with labor, and they maintain control of their product through the production and processing stages. If the farmers and the processor enter the arrangement from positions of equality, the processor can also benefit by reducing the risk of having insufficient inputs to operate the processing facility at its optimum level.

As Richard Levins (2001) argues, farmers at the production level are the only place where competition still exists in agricultural production since input and processing markets are highly concentrated, as are food manufacturing and retail markets (Hendrickson and Heffernan 2002). Levins suggests that farmers need to quit competing among themselves and instead organize to negotiate from a position of power for better prices. The same principle should exist when farmers of the middle think of entering into contractual situations. From what position of power are they negotiating vis-à-vis other actors in the chain, and how are they going to even out that power? Fair trade networks are one example that farmers of the middle can look to for ideas on how to do this.

A third option is to become part of the evolving vertically integrated system as essentially a laborer. In this arena, policy becomes critical in determining how the market operates (see chapter 12, this volume). As Rick Welsh and colleagues (2001) demonstrated, anticorporate farming laws enacted at the state level can influence the structure of agriculture in particular locales. In the last five years, legislative campaigns in over fifteen states attempted to enact a Producer Protection Act (Miller et al. 2000) commonly referred to among activists as the "Contract Grower's Bill of Rights," elements of which are contained in the Fair Contracts for Growers Act of 2003 (S. 91) introduced in the U.S. Senate.

This act can be seen as a way to even out the power positions in production contracting by attempting to give agricultural growers some of the same contractual rights as those that exist under consumer law, plus other protections. It defines retaliation or discrimination against producers who exercise certain rights (such as the right of producers to join producer organizations) as an unfair practice. The Producer Protection Act highlights the political nature of production contracts, and points to the importance of not only economic concerns but also issues of human rights and social justice. Given that agribusiness firms are in a more powerful bargaining position than contract producers, there is a need for such legislation to regulate contract negotiations in ways that will protect producers from possible abuses.

Enforceable competition policy could also give producers a better position of power from which to negotiate contracts. As shown above, increasingly consolidated markets for most agricultural commodities guarantee that farmers have little choice about what to produce or little control over production given the limited marketing options available. Enacting policy reforms that would encourage greater competition in commodity markets could go far to level out the imbalance of power between farmers of the middle and the firms from which they buy or to which they sell.

In summary, farmers of the middle must remember that poultry contracts in Louisiana were economically satisfactory until the broiler market consolidated and there were fewer integrating firms willing to offer contracts. Farmers lost their individual choice and consequently their power. If farmers of the middle choose to enter into contracts in newly emerging short supply chains, they have to think collectively about how they will retain their power position in the contract. This raises some key questions. What happens when other farmers can produce the same innovative products that started the chain, and offer an alternative from which food processors or retailers can source? What happens if the market niche passes? How will farmers retain their control in the supply chain and thus some of their power? Levins (2001) consistently suggests that recognizing the unique attributes of one's valuable product and acting collectively to negotiate together is the only answer to retaining power—and thus price benefits.

To retain their power, farmers of the middle would do well to look at the lessons learned from poultry contracting in Union Parish. As we have discussed, contracts entered into by two principals (parties) on an equal footing can be beneficial in reducing transaction costs and uncertainty. Yet when one principal gains the upper hand, either through the accumulation and exercise of market power, or the elimination of market access, contracts become a way of further reducing farmer autonomy and farm-level profits. Hence, it is not the contract itself that is good or bad, it is simply the structure of agriculture and the community context in which it arises that can have negative social, economic, and ecological consequences. The necessity for negotiating from equal positions of power requires policies that encourage competition in agricultural markets and ameliorate the effects of market power in contracting.

Notes

We would like to thank the USDA Southern Sustainable Agriculture Research and Education Program, the University of Missouri Agricultural Experiment Station, and the Louisiana State University Agricultural Experiment Station for providing funding for various stages of this longitudinal study.

1. They describe four mechanisms: measurement, input control, monitoring, and residual claimancy.

2. During the research period, in Sabine Parish, Louisiana, several growers filed a lawsuit against ConAgra claiming that the firm asked them to make technological changes to experiment with growing bigger birds. After one flock of chickens, the growers alleged, ConAgra backed out of the deal because it did not appear to be profitable (Krebs 2000).

3. This trend held true for early turkey and pork production contracts in Missouri as well.

4. Even here the question is: Which party is most eager to enter the arrangement? A farmer facing major financial problems may see the production contract as their only chance for survival and be willing to sign a contract that is less favorable than a farmer who sees other options.

5. Lane Poultry was the largest broiler producer and processor in the United States following its purchase of Valmac Industries in 1980, but it was still a single-product producer. Lane lost Valmac Industries and then was itself purchased by Tyson Foods because of its economic losses in nine of eleven consecutive quarters in the late 1970s and early 1980s.

References

Ellerman, David. 1992. *Property and Contract in Economics: The Case for Economic Democracy*. Cambridge, MA: Blackwell.

Fox, Alan. 1974. *Beyond Contract: Work, Power, and Trust Relations*. London: Faber.

Harvey, David. 1990. *The Condition of Postmodernity*. Oxford: Basil Blackwell.

Heffernan, William D. 1984. Constraints in the U.S. Poultry Industry. In *Research in Rural Sociology and Development*, 1:237–260. Greenwich, CT: JAI Press.

Heffernan, William D. 1998. Agriculture and Monopoly Capital. *Monthly Review* 50 (3): 46–59.

Heffernan, William D., Mary Hendrickson, and Robert Gronski. 1999. Consolidation in the Food and Agriculture System. Report to National Farmers Union, Washington, DC, February.

Hendrickson, Mary, and William D. Heffernan. 2002. Concentration of Agricultural Markets. Table prepared for the National Farmers Union, Washington, DC, February.

Hendrickson, Mary, William D. Heffernan, Philip H. Howard, and Judith B. Heffernan. 2001. Consolidation in Food Retailing and Dairy. *British Food Journal* 3 (10): 715–728.

Johnson, Jim, Mitch Morehart, Janet Perry, and David Banker. 1996. Farmers' Use of Marketing and Production Contracts. Agricultural Economics Report no. 747. Washington, DC: U.S. Department of Agriculture, Farm Business Economic Branch, Rural Economy Divison, Economic Research Service.

Krebs, Albert. 2000. Louisiana Poultry Growers: "It Was Either Play with ConAgra's Ball or Don't Play." *Agribusiness Examiner*, no. 86, August 24.

Levins, Richard A. 2001. *An Essay on Farm Income*. Staff paper P01–1. University of Minnesota, Department of Applied Economics, College of Agricultural, Food, and Environmental Sciences.

Marsden, Terry, Jo Banks, and Gillian Bristow. 2000. Food Supply Chain Approaches: Exploring Their Role in Rural Development. *Sociologia Ruralis* 40 (4): 424–438.

McMichael, Philip. 1999. Virtual Capitalism and Agri-Food Restructuring. In *Restructuring Global and Regional Agricultures: Transformations in Australasian Agri-Food Economies and Spaces*, ed. David Burch, Jaspar Goss, and Geoff Lawrence, 3–22. Aldershot, UK: Ashgate.

Miller, Thomas J., Heidi Heitkamp, Ken Salazar, Karen Freeman-Wilson, Albert B. Chandler III, Mike Hatch, Mike Moore, Jeremiah W. (Jay) Nixon, Joseph P. Mazurek, Don Stenberg, Frankie Sue Del Papa, W. A. Drew Edmondson, William H. Sorrell, Darrell V. McGraw Jr., James E. Doyle, and Gay Woodhouse. 2000. Statement of State Attorneys General on "Producer Protection Act." September 13. Available at ⟨http://www.state.ia.us/government/ag/agcontractingstatement.htm⟩ (accessed July 11, 2004).

Perry, Janet, David Banker, and Robert Green. 1999. *Broiler Farms' Organization, Management, and Performance*. U.S. Department of Agriculture, Research Economics Division, Economic Research Service.

Teeven, Kevin. 1990. *A History of the Anglo-American Common Law of Contract*. New York: Greenwood.

U.S. Department of Agriculture (USDA). 2002. Census of Agriculture, Louisiana. Washington, DC: National Agricultural Statistics Service.

Wallerstein, Immanuel. 2000. From Sociology to Historical Social Science: Prospects and Obstacles. *British Journal of Sociology* 51 (1): 25–35.

Watts, Michael J. 1994a. Epilogue: Contracting, Social Labor, and Agrarian Transitions. In *Living under Contract: Contract Farming and Agrarian Transformation in Sub-Saharan Africa*, ed. Peter D. Little and Michael J. Watts, 248–257. Madison: University of Wisconsin Press.

Watts, Michael. 1994b. Life under Contract: Contract Farming, Agrarian Restructuring, and Flexible Accumulation. In *Living under Contract: Contract Farming and Agrarian Transformation in Sub-Saharan Africa*, ed. Peter D. Little and Michael J. Watts, 21–77. Madison: University of Wisconsin Press.

Welsh, Rick. 1998. The Importance of Ownership Arrangements in U.S. Agriculture. *Rural Sociology* 63 (2): 199–213.

Welsh, Rick, Chantal Line Carpentier, and Bryan Hubbell. 2001. On the Effectiveness of State Anti-Corporate Farming Laws in the United States. *Food Policy* 26 (5): 543–548.

Williamson, Oliver E. 1991. The Logic of Economic Organization. In *The Nature of the Firm: Origins, Evolution, and Development*, ed. Oliver E. Williamson and Sidney G. Winter, 90–116. Oxford: Oxford University Press.

Wolf, Steven, Brent Hueth, and Ethan Ligon. 2001. Policing Mechanisms in Agricultural Contracts. *Rural Sociology* 66 (3): 359–381.

III

Bringing Mid-Level Supply Chains and Consumers Together

6

Consumer Considerations and the Agriculture of the Middle

Eileen Brady and Caitlin O'Brady

An apple is an apple is an apple. Or is it? This question and others like it are on the lips of many U.S. consumers. The food and farming industries are slowly waking up to a twenty-first-century reality that the increasingly globalized food system, while providing inexpensive products, does not address many of the latest food safety and health concerns of consumers. Nor does it satisfy an underlying desire for the unique tastes, flavors, and pride of place that food can offer. Many shoppers, having lost confidence and trust in the large food purveyors, are asking where their food comes from, how it was grown or manufactured, and who produced it. At the same time they are seeking foods that are fun, delicious, and come with a story. While many analysts in the food industry have argued that consumers with these types of preferences are few, isolated, and do not represent a substantial market opportunity for growers and purveyors of quality and local food, the evidence is mounting that the market opportunity is more significant than initially thought.

These same analysts have outlined the battle lines for the food industry. Much is being made of the turf being carved up for what some call the "mass food" consumers and the "class food" consumers. The conversation at conventional food conferences and trade shows eventually turns to Wal-Mart, the predominant provider of mass food—basic foods for the bulk of the population at an affordable price. Wal-Mart, the giant in the industry, with revenues of over $250 billion in 2003 and almost five thousand stores (*Chain Store Age* 2004), is expected to dominate the market for basic foodstuffs for the foreseeable future. Many other large grocery chains and quick-service restaurants are fighting to maintain their share of the mass food dollar. The smaller, but iconic, Whole Foods Market chain has been given the mantle of the premier supplier of class food to those who can afford it. With almost $3 billion in revenues in 2003 and 145 stores, Whole Foods Market is continually resetting the bar in the quality foods market (*Chain Store Age* 2004). Farmers' markets and many top-end food service operations also lay claim to servicing the class food consumers.

The truth, however, is murkier than the schism portrayed by the analysts of a growing division in the population between those who purchase cheap, processed foods and those who can afford healthy, quality foods. There is mounting evidence that a fundamental shift toward fresh and freshly prepared quality foods made from ingredients produced by family farmers, many of whom are local to the region, is on the rise across many segments of the population. (See the consumer studies available at ⟨http://www.farmprofitability.org/research.htm⟩.)

An apple is not just an apple to these populations. Apples grown without harmful chemicals, apples grown by family farmers, and specialty varieties of apples grown within the eater's home region—these are the apples to satisfy this growing consumer base. Producers, manufacturers, retailers, and restaurateurs take heed. A consumer base is building for local and quality foods. Midsize growers who can come to understand these consumers and market effectively to them may find a new direction for their farm operations.

When Did "Food Scare" Become Part of Our Vocabulary?

Nineteen fifty-nine ushered in the first modern-day "food panic," as it was called then. That year, on the day before Thanksgiving, the secretary of health, education, and welfare sounded the alarm that cranberries may have been "contaminated" with the weed killer aminotriazole. Consumers reacted immediately, refusing to buy cranberries. Some states banned cranberry sales. Many schools and restaurants refused to serve them for years. The cranberry farmers struggled for a decade to regain a footing in the marketplace, though as it turned out, only trace amounts of the weed killer were found in the berries (Whelan 2004).

Since the 1980s, food scares have escalated in number, slowly undermining consumer confidence. In 1988, salmonella was found in eggs across the United Kingdom, forcing the slaughter of four million hens and four hundred million eggs (Pennington Group 1997). The alar apple scare struck the United States in 1989 when *60 Minutes* aired a segment about a report from the Natural Resources Defense Council that claimed alar was a carcinogen to be taken seriously. Used to prevent apples from dropping early, alar was linked to lung and kidney tumors in mice. The apple industry unsuccessfully sued *60 Minutes*, claiming a loss of $100 million. Publicly reported E. coli outbreaks surged in the 1990s (Reilly et al. 2001). E. coli, an infectious bacteria found mostly in beef products, was to blame for four deaths from meals consumed at Jack in the Box quick-service restaurants in 1993 and at least eighteen deaths in 1996 from meat purchased from a butcher shop in Lanarkshire, Scotland (Reilly 2001).

More recently, mad cow disease, or bovine spongiform encephalopathy (BSE), was found in over a hundred thousand cattle in the 1980s and 1990s (World Organization for Animal Health 2005). The disease was subsequently discovered to be potentially linked to a deadly human illness, Creutzfeldt-Jakob disease. Over four million cattle were slaughtered in the United Kingdom, and when the first diagnosis of the disease was found in the United States, beef prices tumbled across the nation. Even milk, seemingly the most wholesome food product of all, has not escaped safety-related questions. The genetically engineered recombinant bovine growth hormone, used to increase milk production in dairy herds, has been linked to increases in breast and prostate cancer. Canada, Australia, New Zealand, and the European Union have banned the use of the hormone (Rebbins 2005). In northern California, most dairies, reacting to consumer demand, have discontinued the hormone's use (North et al. 2004).

While industry groups have spent millions on damage control to mitigate food scares, an underlying impression of concern and caution about food safety in the industrial food industries has begun to surface. Epitomized by the stories told in *Fast Food Nation*, a best seller by Eric Schlosser (2001), the imperfections in the globalized system of mass-produced foods are becoming apparent to the average consumer. Not budging from the *New York Times* best-seller list for eighty-five weeks, the book hit a nerve, selling over a million copies in six printings.

In addition, a new and intensifying concern about food safety as a result of a potential terrorist attack in the United States focuses on what has been dubbed "agro-security." In April 2004, reacting to pressures to prepare for possible national security events, the U.S. government deemed this issue worthy of study, providing $33 million in grants to the University of Minnesota and Texas A&M University to outline a food and water protection agenda (Department of Homeland Security 2004). The once-trusted U.S. food safety program, circa 1950, is simply not sufficient to provide complete confidence for the consumer of 2005. Until food safety concerns are addressed by new products and practices that engender greater trust, they appear to be part of our future.

These reactions to food scares come from more than a few niche consumer segments. Millions of dollars in lost sales, lost product, and the retooling of old practices due to food supply panics are signs of widespread concern and loss of trust in the conventional food delivery system. In a nationwide survey conducted by Roper Public Agencies for Organic Valley Family of Farms in 2004, only 24 percent of U.S. consumers place "a lot of trust" in large-scale industrial farms, and by a 71 to 15 percent margin, believe that smaller-scale family farms are more likely to care about food safety than large-scale industrial ones (McGovern 2004). Growers providing transparent methodologies that offer assurances for the twenty-first-century consumer will be well positioned to reap the benefits of a growing market need for a sophisticated food safety promise.

An Apple a Day Isn't Enough: Obesity and Public Health

Obesity. A one-word wake-up call. In 2001, a report titled *The Surgeon General's Call to Action to Prevent and Decrease Overweight and Obesity* stimulated media concern, reaction from top-level businesses, further studies, and a myriad of proposed solutions. The report found that approximately two-thirds of the adult population in the United States is overweight or obese. "Overweight and obesity may soon cause as much preventable disease and death as cigarette smoking," said Surgeon General David Satcher (2001). A burgeoning collection of media stories and reports have piled up in the years since Satcher's report detailing how obesity contributes to chronic diseases including diabetes, hypertension, stroke, cardiovascular disease, and some cancers. The situation has been called "dire" and labeled an "epidemic." Indeed, in 2004, then Surgeon General Richard H. Carmona asserted, "We may see the first generation that will be less healthy and have a shorter life expectancy than their parents."

Top-level public policy conversations have ensued. The priorities of the 2005 USDA food guidelines and pyramid review committee are clear from its executive summary: "The Committee was especially interested in finding strong scientific support for dietary and physical activity measures that could reduce the Nation's major diet-related health problems —overweight and obesity, hypertension, abnormal blood lipids, diabetes, coronary heart disease (CHD), certain types of cancer, and osteoporosis" (U.S. Department of Agriculture 2005). The pyramid is expected to be thoroughly critiqued for its ability to address the spiraling diet-related health problems of the current generation.

Internationally, the World Health Organization denounced the U.S. diet in 2003, with direct recommendations to reduce the intake of processed foods along with foods high in salt and sugars. The 2005 World Health Congress, generally keynoted by speakers with political affiliations such as Robert Kennedy, invited John Mackey, chair of Whole Foods, to speak on how best to link food choices and health in consumers' minds. The World Health Organization (2004) is pressing for market-based guidance for solving the health issues linked to the U.S. diet.

Today, topics of obesity and food are of interest to many people, and have inspired an odd sort of popular entertainment. In 2003, an independent documentary made for a mere $65,000 won the Sundance film festival's best director award and caught the attention of a nation. *Supersize Me*, a film that asks the question "What would happen if you ate nothing but fast food for breakfast, lunch, and dinner?" grossed over $11 million at the box office in its first five months (MovieWeb 2005). The protagonist in the film eats McDonald's meals for thirty days, gaining a significant amount of weight and causing notable damage to his health in the short period. Moviegoers watch in awe as doctors express extreme concern about the self-induced twenty-four-hour fast-food approach to eating.

Whether by noting their own waistlines or wading through the enormous numbers of studies available on the topic, consumers are coming to terms with the fact that what they are eating may be part of the problem—and part of the solution. In fact, an increasing number of consumers are looking at food as having medical benefits. Over 50 percent think some medical therapies and drug use can be reduced by eating certain foods—up almost 10 percent from 1994 (Stark 2001).

Switching to a USDA-recommended diet of five fruits and vegetables a day, or a cancer-prevention diet of seven servings of fruits and vegetables a day is a healthy program for eaters, and could give farmers an enormous boost. A study conducted by Karen Jetter and colleagues (2004) at the University of California Agricultural Issues Center estimates that a shift to recommended diets would create an estimated annual net nationwide benefit for fruit and vegetable farmers that ranges from $460 million to $1.44 billion. Farmers, fishers, and ranchers listening closely to the U.S. consumer, and willing to alter their production plans to meet the needs of the health shopper, will be positioned for future growth and survival.

Home Is Where the Heirloom Tomatoes Are

The globalized food and agriculture system has brought consumers convenience, low prices, and a predictable and familiar selection of foods, but it has also contributed to the loss of a sense of place. No matter where you travel in the United States, McDonald's is ready to greet you, the Olive Garden is prepared to entertain you, or Wal-Mart will be happy to fulfill your shopping list. While the comfort and confidence in recognizable brands and stores has been welcomed by much of the United States, the bill for it is now past due. Communities have lost both regional identities and the unique qualities of place that residents can take pride in. From Cleveland, Ohio, to Fort Worth, Texas, cities and towns are struggling to attract jobs, capital, and talent, and are striving to present themselves to businesses considering a move to a place offering a premier quality of life. This has become more difficult as a global, one-size-fits-all culture is taking root. Cities are stretching to market their town as having a unique identity.

Nowhere is the development of mass culture more apparent than in the foods we eat. Boston's baked beans, clam chowder, blueberry pie, and johnnycakes have helped define home for thousands of people in the Northeast. Wild salmon has been an icon of the Pacific Northwest for centuries. These foods, and others like them, rather than being utilized to promote the pride and economic attractiveness that comes with distinctive regional identities, have been overwhelmed by the sheer amount of mass-produced foods. Regional foods and local farms, once known by urban dwellers, no longer play a large role in characterizing the regions of the United States.

To produce and distribute foods efficiently around the world, the food and farming industries have had to narrow their fresh offerings to products that can be produced uniformly and transported easily. Hundreds of varieties of heirloom tomatoes, melons, strawberries, pork, poultry, and others have been discontinued and even lost over the years. They simply are not shelf stable and hardy enough to travel long distances.

As these systems of efficiency were developed, markets for local farm-fresh products were lost. Ken Meter and Jon Rosales (2001)—respectively, from the Crossroads Resource Center and the Institute for Social, Economic, and Ecological Sustainability—detail the lost local market opportunities for farming communities in southeastern Minnesota in their report *Finding Food in Farm Country*. The 8,436 farms in the area sold $866 million worth of farm products in 1997. Meanwhile the 303,256 residents in the region purchased $506 million in food, mostly from producers outside the state. If the $400 million spent by farm families for inputs and supplies that come from outside the region are figured into the equation, it becomes obvious that millions of dollars are flowing out of the community instead of building wealth in the area (Meter and Rosales 2001). Additional studies in California have shown a similar trend.

But consumers are beginning to question some aspects of the global food system, demanding the sweeter, diverse tastes that regional foods provide. As Brian Halweil says in the World Watch book *Eat Here* (2004), "There is an unavoidable tension between the human enjoyment of variety and the global homogenization of food." The experience of this tension may be the early signs of consumer demand for regional harvests.

It is not just an indulgence in the tastes of regional foods that entreat these consumers. In the past decade, consumers have shown in small yet substantial ways an interest in rebuilding urban-rural relationships and getting to know the farms that surround their communities. According to the USDA, the number of farmers' markets in the United States has grown dramatically, increasing 79 percent from 1994 to 2002 (Agricultural Marketing Service 2002). Likewise, the CSA movement that began in the United States in 1986 now numbers over one thousand farms (Lass et al. 2003). In a CSA, a consumer buys a "share" of the farm goods expected for the year, participates in farm workdays, and generally gets to know the farmers who provide their weekly share.

Fast Company announced that "local is the new organic," charging that the latest food trend to emerge is a preference for local buying (Lidsky 2004). Some markets are providing "buy local" signage that identifies the farmer and tells the story of the farm associated with each product. Many restaurants list the names of the local farms that supply the ingredients for the menu items. Buy Fresh, Buy Local campaigns have sprung up around the country. At the New Leaf Grocery chain on the central coast of California, produce signs complete with photos of farmers don the stores. When given a choice, as in the 2002–2003 Buy Local tomato campaign in Portland, Oregon, championed by several retailers, consumers will decide to purchase local produce. The first year of this project saw a 40 percent increase in the purchase of local tomatoes over unidentified ones, even though price of the locally grown tomatoes was in some cases higher (Berkenkamp 2003).

The Community Food Security Coalition, launched in 1995 to promote solutions to hunger and the globalization of the food system and in 2005 numbered six hundred organizational members, reveals a growing interest in building communities focused on regional food economies. The coalition has worked successfully to enact federal legislation to support farm-to-cafeteria projects (see ⟨http://www.foodsecurity.org⟩). The Farm-to-Cafeteria Projects Act was designed to provide schools and nonprofits with seed grants to cover the initial costs of preparing locally grown food for school meals. The bill passed, and, as of this writing, awaits funding allocations and enactment.

Driven by consumer and civic interests, over fifty food policy councils have been created across the country since 1997 (Borren 2003). In pursuit of planning for food security and the promotion of local foods, these councils are beginning to be heard. They are proposing local buying preferences for county prisons and public schools, securing permanent sites for farmers' markets, and lobbying for farmland inside the city, known as urban agriculture.

This grassroots local foods movement offers an opportunity for rural and coastal growers to reengage with urban consumers and develop new markets close to home. After years of divisiveness between urban and rural communities, there is indication that the time is ripe to renegotiate the social contract between them. Urban eaters are rediscovering the value of local foods and the benefits that come from supporting local farming communities. The interest from consumers in community-based local food systems, while nascent, is growing more sophisticated, and is arming advocates with arguments for reorienting the food-buying habits of retailers, restaurants, and public institutions. *Ripe for Change: Rethinking California's Food Economy* (Mamen et al. 2004) is the most recent handbook for advocates of local foods that outlines macroeconomic rationales for community-based foods. The reasons include existing redundant interstate trade, how consolidation in the food industry has squelched local food initiatives, and the disproportionately small percent of retail dollar that growers receive. If growers can reach across to build relationships with the local food movements that are growing in neighboring urban areas, opportunities for building healthy, economically viable urban-rural partnerships may be harnessed.

Who Is Responding to New Consumer Demands?

What's more American than hot dogs at the ballpark? This favorite is making way for a new addition: the organic hot dog. Responding to consumer demand, San Diego's Petco Park and Saint Louis' Busch Stadium tested selling certified organic hot dogs and bratwurst during the 2005 season (Horovitz 2004). The organic dogs were sold for four dollars each—a dollar more than their conventional counterparts. Mainstream purchasers across the United States, from ballparks to pizza shops, are testing these emerging market trends. In fact, McDonald's is a 90 percent owner of a new fast-food operation—Chipotle Mexican Grill—that features natural meats (Murray 2005) and boasts some four hundred eateries in the United States. Though Chipotle raised its prices to cover higher costs, its sales increased tenfold (Huynh 2003).

While the natural food and product grocery industry is growing at an envious pace, traditional supermarkets have countered by starting to bring in healthy, organic, and regionally known products. In fact, 2002 was the first year more natural food products were sold in conventional grocery stores than natural food stores. As the National Grocery Association notes, "Natural foods are the fastest growing product area in the supermarket today. For retailers, wholesalers and manufacturers the question is not whether this trend will continue, but how to seize the opportunity natural and organic foods present" (Seltzer 2004). The association encourages its member stores to consider adding natural food items or even whole natural product sections. Wal-Mart's plan to add organic products underscores how seriously buyers are taking this market opportunity. No longer are natural foods found only in crunchy-granola, health food stores. They are positioned to be the next big growth area for many traditional operations.

Even Kraft Foods (2003), pushed by obesity concerns, announced a healthy-lifestyle initiative, a commitment to retool products, and the formation of a global advisory group to provide direction on health and nutrition. The larger manufactures have seen an enormous market opportunity in the natural and healthy products industry—so much so that many of the small, natural food brands have been purchased by industry consolidators. Small Planet Foods, originally known only to customers of small health food stores, was picked up by General Mills in 2000. Now General Mills is actively expanding the market for the Small Planet Foods brands—Cascadian Farm and Muir Glen. Horizon Organic, a growing organic milk and dairy products manufacturer, was purchased in 2003 by Dean Foods Company, the leading U.S. producer of milk and dairy products.

Food service leaders like SYSCO and Bon Appétit have jumped into the quality foods market as well. Rick Schnieders, CEO for SYSCO, a $23 billion corporation in 2003, speaks regularly with family farmers to explain the mounting consumer demand for farm-fresh products. Providing foods for restaurants and cafeterias nationwide, SYSCO directs purchasers to meet the needs of those operations that are asking for regionally produced, fresh foods. SYSCO is also investigating environmentally sound agricultural practices and integrated pest management systems for their growers to employ to meet the anticipated demand for these types of products (Schnieders 2004).

Bon Appétit, a private on-site restaurant management company, with a client list that includes Intel, Yahoo!, Nordstroms, American University, and Massachusetts Institute of Technology, has organized its operations so that chefs in each district can make regional purchasing decisions and are not bound by national buying strategies. The quality product offerings that Bon Appétit has can be directly traced to its commitment to the local purchasing of ecologically grown foods. One of the key stated beliefs of the organization clearly defines its commitment to quality and social responsibility: "Buying local and sustainable ingredients preserves flavor and regional diversity while investing in the community" (Bon Appétit 2005).

The national Chefs Collaborative and the Slow Food movement have both mobilized thousands of chefs around the country to take up the banner of delicious, local, and seasonal foods. Renowned chef Alice Waters (2005), often billed as the "grandmother of the seasonal foods movement," opened her restaurant Chez Panisse in Berkeley, California, in 1971. The upscale eatery is known for its stated belief that "the best-tasting food is organically grown and harvested in ways that are ecologically sound, by people who are taking care of the land for future generations." Reservations are required long in advance for a seat at this famed dining establishment. Waters and other white tablecloth restaurants have been on the leading edge of reviving an almost-lost recognition of the delights brought by eating foods at the height of their season.

Nevertheless, the myth that the upper-end restaurants have cornered the market for local and quality foods is just that. More and more mainstream food service operations have also moved in to service the quality and local food customer. Tod Murphy, founder of the Farmers Diner in Quechee, Vermont (⟨http://www.farmersdiner.com⟩), offers a $6.95 breakfast of home fries, eggs, sausage, and toast. This doesn't sound like a big deal until you realize the potatoes are from a vegetable grower down the road; the eggs are from a poultry farm in New Haven, Vermont; and the sausage comes from Vermont-raised chicken and pork. "You get a wholesome meal and a chance to support local farmers," says Murphy (2004).

Across the country in Portland, Oregon, Hot Lips Pizza (⟨http://www.hotlipspizza.com⟩) is putting a new face on pizza. The names and stories of the local farmers who provide the wheat for the crust and the toppings stacked on top are prominent in the four pie shops in the area. Likewise, Burgerville, based in Vancouver, Washington (⟨http://

www.burgerville.com⟩), with thirty-nine quick-service restaurants in Washington and Oregon, prides itself on a top-quality, localized menu. A commitment to naturally raised Oregon Country Beef (⟨http://www .oregoncountrybeef.com⟩) and huckleberry shakes when the berries are in season are signature items. And one of this chapter's authors and her husband as well as two other families, along with about fifty friends, founded a grocery store in Portland, Oregon, focused on providing sustainable, regionally grown products. New Seasons Market (⟨http://www.newseasonsmarket.com⟩), a nine-store chain, is committed to building the health of the community, strong urban-rural partnerships between producers and consumers, and showcasing sustainably produced produce, meats, fish, and manufactured products.

While catering to the growing demands for safe, healthy, and regionally produced foods may have once seemed like a small niche industry, the consumer interest is beginning to show signs of coalescing as a serious market trend, and in turn, providing an opportunity for enterprising farmers who are willing to learn and service this new and growing consumer base.

Reconnecting the Farmer to the Market

The consumers of this emerging market want to know more about where their food comes from. They want assurances that their food choices are healthy and safe. They want the story behind the product. And they want accountability. Who produced the food they are eating? How was it grown? What's the name of the farmer? The new farmer, rancher, or fisher must realize they are selling more than the product they are producing. They are building a new relationship—one that requires participation and interest from the producer as well as the consumer. Producers must reach out and understand their consumers. If a farmer's harvest is given to a broker and changes hands several times, the grower never knows where the product ends up, and the relationship with the consumer is severed. Producers must know where their products end up. They must come to know the end consumers along with their likes, needs, and wants. Products must be adjusted to meet these needs, building on the relationship with the end consumer. Educational information must be constructed to answer consumers' questions and to pique their interest in the product. Growers must consider themselves marketers, always trying to further their knowledge of their customers.

A close connection with the customer is most obviously achieved through marketing directly to the customer—at a farmers' market, for instance. Yet a growing number of producers are seeking to establish this relationship with customers even when they are selling wholesale to a restaurant or grocery. At New Seasons Market, producers regularly come into the store and demo their products on the weekends. This allows them to meet some of the customers and listen closely to the consumer reaction to their products. Some of the ranchers from whom New Seasons Market buys spend time working behind the meat counter, again getting to know how their product presents to shoppers and how it is received.

There are many ways to begin understanding the market for your products. Producers might consider subscribing to the health and natural products industry magazines: *LOHAS*, *Natural Foods Merchandiser*, and *Supermarket News*. Farmers, fishers, and ranchers can join food policy councils, which are often struggling to get producer participation. Attending the Community Food Security Coalition conferences and events can offer knowledge about community-based grassroots efforts. Most important, growers should not be satisfied with just knowing their broker. Know the buyer behind the broker. Know the customer behind the buyer. Visit places that the product is sold. Talk to the customer.

References

Agricultural Marketing Service. 2002. *National Farmers Market Directory*. Washington, DC: Agricultural Marketing Service, U.S. Department of Agriculture. Available at ⟨http://www.ams.usda.gov/⟩.

Berkenkamp, Joanne. 2003. *Buy Local Campaign Evaluation*. Portland, OR: Ecotrust, September. Available at ⟨http://www.ecotrust.org⟩.

Bon Appétit Management Company. 2005. Available at ⟨http://www.bamco .com/website/home.html⟩.

Borren, Sarah Marie. 2003. *Food Policy Councils: Practice and Possibility*. Eugene, OR: Congressional Hunger Center, February 12.

Carmona, Richard. 2004. Testimony of Dr. Richard Carmona. Competition, Foreign Commerce and Infrastructure Security hearing, The Rise of Obesity in Children, March 2.

Chain Store Age. 2004. State of the Industry. Available at ⟨http://www .chainstoreage.com/archives/preview.cfm?ID=2004214221489&SC=state+of+the +industry&CFID=328417&CFTOKEN=17255791⟩.

Department of Homeland Security Press Office. 2004. *Homeland Security Selects Texas A&M University and University of Minnesota to Lead New Centers of Excellence on Agro-Security.* Washington, DC: U.S. Department of Homeland Security, April. Available at ⟨http://www.dhs.gov/dhspublic/ display?theme=43&content=3515&print=true⟩.

Halweil, Brian. 2004. *Eat Here: Reclaiming Homegrown Pleasures in a Global Supermarket.* Washington, DC: World Watch Institute. Available at ⟨http:// www.worldwatch.org/⟩.

Horovitz, Bruce. 2004. Get 'em While They're Hot: Stadium Franks Go Organic. *USA Today,* December 1.

Huynh, Dai. 2003. US Restaurants Serving More Natural and Organic Meats. *Houston Chronicle,* October 22.

Jetter, Karen M., James A. Chalfant, and Daniel A. Sumner. 2004. Does 5-a-Day Pay? *ACI Issues Briefs* 27 (September). Agricultural Issues Center, University of California at Davis. Available at ⟨http://aic.ucdavis.edu/pub/briefs/brief27.pdf⟩.

Kraft Foods. 2003. *Kraft Foods Announces Global Initiatives to Help Address Rise in Obesity.* Northfield, IL: Kraft Foods, July. Available at ⟨http://www .kraft.com/newsroom/07012003.html⟩.

Lass, Daniel, G. W. Stevenson, John Hendrickson, and Kathy Ruhf. 2003. *CSA across the Nation: Findings from the 1999 CSA Survey.* Madison: Center for Integrated Agricultural Systems, University of Wisconsin.

Lidsky, David. 2004. Fast Forward 2005. *Fast Company* 88 (November): 69. Available at ⟨http://www.fastcompany.com/magazine/88/fast-forward-12-17 .html#fastforward12⟩.

Mamen, Katy, et al. 2004. *Ripe for Change: Rethinking California's Food Economy.* Berkeley, CA: International Society for Ecology and Culture. Available at ⟨http://www.isec.org.uk/articles/RipeForChange.pdf⟩.

McGovern, Sue. 2004. *Roper Food and Farming Survey.* La Farge, WI: Organic Valley, April. Available at ⟨http://www.organicvalley.com/⟩.

Meter, Ken, and Jon Rosales. 2001. *Finding Food in Farm Country.* Lanesboro, MN: Community Design Center. Published in cooperation with Crossroads Resource Center and the University of Minnesota. Available at ⟨http://www .crcworks.org/ff.pdf⟩.

MovieWeb. 2005. Box Office: *Supersize Me.* May 27. Available at ⟨http:// www.movieweb.com/movies/box_office/daily/film_daily.php?id=2421⟩.

Murphy, Tod. 2004. Personal communication.

Murray, Barbara. 2005. *Chipolte Mexican Grill, Inc.* Austin, TX: Hoovers, Inc.

North, Rick, et al. 2004. rBGH-Free Oregon Campaign Fact Sheet. Portland: Physicians for Social Responsibility, Oregon Chapter.

Pennington Group. 1997. *Report on the Circumstances Leading to the 1996 Outbreak of Infection with E. coli O157 in Central Scotland, the Implications for Food Safety, and the Lessons to Be Learned.* Edinburgh: Stationery Office.

Rebbins, Lehn. 2005. Is rBST the Same as rBGH? In *The Food Revolution.* Available at ⟨http://www.foodrevolution.org⟩.

Reilly, Bill, et al. 2001. *Task Force on E. coli O157—Final Report.* Joint Food Standards Agency Scotland and Scottish Executive Task Force on E. coli O157, Edinburgh. Available at ⟨http://www.food.gov.uk/multimedia/pdfs/ecolitaskfinreport⟩.

Satcher, David. 2001. *The Surgeon General's Call to Action to Prevent and Decrease Overweight and Obesity.* Washington, DC: U.S. Department of Health and Human Services, December. Available at ⟨http://www.surgeongeneral.gov/topics/obesity/calltoaction/toc.htm⟩.

Schlosser, Eric. 2001. *Fast Food Nation: The Dark Side of the All-American Meal.* Boston: Houghton Mifflin.

Schnieders, Rick. 2004. Personal communication.

Seltzer, Jonathan M. 2004. Natural Foods: A Natural Profit Opportunity. *National Grocer.* Arlington, VA: National Grocer Association. Available at ⟨http://www.nationalgrocers.org/NGNaturalFoods.html⟩.

Stark, Myra. 2001. *The State of the U.S. Consumer 2001.* New York: Saatchi and Saatchi.

U.S. Department of Agriculture and Health and Human Services. 2005. Executive Summary. *Dietary Guidelines for Americans 2005.* January. Available at ⟨http://www.health.gov/dietaryguidelines/dga2005/document/html/executivesummary.htm⟩.

Waters, Alice. 2005. About Chez Panisse. Available at ⟨http://www.chezpanisse.com/glance.html⟩.

Whelan, Elizabeth. 2004. *Remembering a Berry Scary Thanksgiving (from Tech Central Station).* American Council on Science and Health, New York. Available at ⟨http://www.acsh.org/news/newsID.1002/news_detail.asp⟩.

World Health Organization. 2004. *Strategy on Diet, Physical Activity, and Health.* Fifty-seventh world health assembly, resolution WHA57.17. Available at ⟨http://www.who.int/gb/ebwha/pdf_files/WHA57/A57_R17-en.pdf⟩.

World Organization for Animal Health. 2005. *Number of Reported Cases of Bovine Spongiform Encephalopathy (BSE) in Farmed Cattle Worldwide.* May. Available at ⟨http://www.oie.int/eng/info/en_esbmonde.htm⟩.

7

Values-Based Supply Chains: Strategies for Agrifood Enterprises of the Middle

G. W. Stevenson and Rich Pirog

Caught in the middle as the U.S. food system divides into global commodity marketing, on the one side, and the direct marketing of food to local consumers, on the other, many traditional family farms across the country are increasingly at risk (see chapter 1). A significant part of the peril is the result of the increasing concentration in the processing and retail sectors of the system that creates power imbalances in market relationships. Such imbalances enable strategic behavior in traditional agrifood supply chains that often seriously disadvantage the least powerful participants, notably farmers, ranchers, and other food enterprises in the middle like regionally based food processors, distributors, and retailers (see chapters 5 and 12). Restoring balance to these agrifood economic relationships will require changes in both private-sector business models and public policy (see chapter 8). This chapter explores one strategy for such new business models: values-based supply chains, or more succinctly, value chains.

Shifts are occurring in the food system as well as the larger social economy that can provide significant opportunities to develop farming and food systems in which a reformed agriculture of the middle can prosper. Consumer surveys indicate that a growing number of food buyers are seriously concerned with the freshness and nutritional content of their food, and prefer to purchase food that has been grown locally or regionally on family-scaled farms.[1] Public health professionals are speaking out about the need to address a spectrum of food- and diet-related concerns from antibiotic resistance, through obesity and coronary artery disease, to food-borne illness (Schettler 2004). Following Europe's lead, and emphasizing issues of social justice and environmental responsibility, a growing fair-trade movement has developed in the United States (Jaffee et al. 2004; Raynolds 2000). Finally, progressive leaders in some sizable food corporations are recognizing the confluence of their interests with the maintenance and regeneration of an agriculture of the middle. The CEO of a large food service company describes many of his customers as wanting memorable high-quality food, produced with a farming story they can support, and brought to them through supply chains they can trust (Schnieders 2004).

Value Chains: Potential for Regenerating an Agriculture of the Middle

Value chains are long-term networks of partnering business enterprises working together to maximize value for the partners and end customers of a particular product or service. In the business literature, these long-term interorganizational relationships are also called "extended enterprises," "virtual integration," "strategic alliances," "integrated value systems," and "value-added partnerships" (Dyer 2000, 27; Handfield and Nichols 2002, 5).[2] Value chains are different than many traditional supply chains on important dimensions (van Donkersgoed 2003, 1–2). The key characteristics of value chains include the following:

• Appropriateness for situations in which economies of scale are coupled with complex products that differentiate and add value in the marketplace
• Capacity to combine cooperation with competition to achieve collaborative advantages and adapt relatively quickly to changes in the market
• Emphasis on high levels of performance and high levels of trust throughout the network
• Emphasis on shared vision, shared information (transparency), and shared decision making among the strategic partners
• Commitment to the welfare of all participants in the value chain, including fair profit margins, fair wages, and business agreements of appropriate extended duration

This chapter is particularly interested in what might be called midtier food value chains: strategic alliances between midsize independent (often cooperative) food production, processing, and distribution or retail enterprises that seek to create and retain more value on the front (farmer or rancher) end of the chain, and effectively operate at regional levels.

Economies of Scale and Differentiated Products

The economic sectors where value chains have proven most successful tend to be higher-volume, complex-product industries like the automobile and truck, consumer electronics, or high-end apparel industries (Dyer 2000; Whalen 2001; Fearne et al. 2001). Value chains are less advantage giving in industries where mass commodification strategies predominate like petrochemicals and bulk pharmaceuticals (Dyer 2000, 35). A value chain featuring specialty pork products might include a cooperative of farmer producers at the center, with upstream partners like veterinarians or feed suppliers, and downstream partners like meat packing and fabrication plants, food distributors and marketers, and final consumers of the products.[3]

It will be important to identify the places in the agrifood system where this combination of scale and product differentiation most clearly manifests itself. The initial indicators are that these characteristics describe a growing segment of the food service industry's customer base (Callimich 2004). The demand is increasing for food products with an enlarging range of unique attributes, which may be based on functionality (specialty flours for artisanal bakeries), food safety (antibiotic- and/or hormone-free meat), environmental impact (organic or integrated pest management grown), geographic location (regionally based products with U.S. certification marks, such as Vidalia onions), or other value-giving characteristics (Schnieders 2004). As such markets continue to grow, opportunities will develop for value chains incorporating the country's larger small and midsize farms that have the flexibility to implement innovative agricultural production systems along with the capacity to produce the volumes necessary to supply significant quantities of food into these new food chains (Kirschenmann 2004, 57).

The supermarket sector appears to be a less friendly environment for midtier food value chains, particularly the larger national and international companies like Wal-Mart, Krogers, or Albertsons. Such supermarket chains tend to compete on the volume and price associated with food products manufactured from a few ingredients uniformly produced—notably corn, soybeans, wheat, sugar, and salt (Schnieders 2004). In addition, these supermarkets often have centralized purchasing systems that do not interface well with more regional supply chains (Greenberg 2004, *adversarily · collaborative*

9). Finally, many of these supermarket companies are noted for maintaining adversarial relationships with their suppliers—a different business paradigm than the collaborative approach taken by value chains (Kumar 1996, 93).

Exceptions to the above portrait may be opportunities for value chain collaborations with more regional supermarket chains that are looking for ways to differentiate themselves from their larger national competitors. A good example is New Seasons Market, a regional grocery chain established in 1999 and based in Portland, Oregon.[4] Locally owned and committed to the development of a Northwest regional food economy, New Seasons Market competes successfully with larger chains like Safeway and Albertsons by featuring an affordable selection of quality food products, including both natural and conventional grocery options. Signature foods like lamb, beef, pork, seafood, and organic wheat flours are sourced through agreements with regional producers where price is based on what the producers need and what New Seasons Market can afford. With each party's needs taken into account, the relationships that New Seasons has with its producer partners are described as "longstanding and vital."

Cooperation and Competition

The overall business model of value chains features close cooperation between strategic partners within the chain and competition between chains doing business in given product sectors. Commenting on a sector in which value chains have been highly developed, an analyst of the automobile industry claims that "future competitive advantage will increasingly be created by teams of companies, rather than by single firms. Single firms will simply be unable to amass the resources and knowledge required to compete with a well-coordinated team of competing companies.... Competitive advantage will increasingly be jointly created, and shared, by teams of firms within a value chain" (Dyer 2000, 169).

Value chains contrast with two other business models: arm's-length (market) relationships with suppliers, and vertical integration through the internal ownership of productive capacity. There is a strong case to be made for carefully selecting as value chain members those "strategic partners" whose businesses create the highest value and the greatest differentiation in the marketplace. Other businesses in the supply chain that produce generic or nonstrategic inputs are more appropriately dealt with in some form of arm's-length or market relationship—for example, via online bidding or auctions.[5] Wise value chain construction also involves selecting partners that bring distinctive competencies but similar values and goals (Kumar 1996, 98; Handfield and Nichols 2002, 49).

Studies indicate high-performance scores for value chains in such disparate sectors as apparel, auto and truck, fresh fruits and vegetables, and more general manufacturer-retailer relationships (Whalen 2001, 67; Dyer 2000, chapter 6; Handfield and Nichols 2002, chapter 10; Zuurbier 1999, 27; Kumar 1996, 93). Drawing on their learning and adaptive capacities (Peterson 2002, 1331), value chains are particularly powerful at reducing the costs of product development, production, and procurement transactions as well as increasing the speed to market and overall product quality. As one analyst summarizes the advantages of value chains involving a range of manufacturer-retailer relationships, "Manufacturers and retailers can exploit their complementary skills to reduce transaction costs, adapt quickly to marketplace changes, and develop more creative solutions to meet consumers' needs. Vertically integrated companies are too inflexible and traditional manufacturer-retailer relationships are too adversarial to promote such behavior and skills" (Kumar 1996, 100).

This section has focused on some of the collaborative dimensions of vertical linkages in value chains. An additional dimension with regard to midtier food value chains is the importance of horizontal linkages and collaborations among groups of farmers to assemble sufficient volumes of product to move through significantly scaled food chains. Currently, crucial explorations are occurring with horizontal linkages that take multilateral or bilateral forms (Kaplinsky and Morris 2001, 98). Multilateral linkages describe business organizational forms entered into by farmers or ranchers like cooperatives or limited liability companies where decision making is done collectively. Examples of successful cooperatives of the middle include Oregon Country Natural Beef in the Pacific Northwest and Ozark Mountain Pork in Missouri (antibiotic- and hormone-free meat), Thumb Oilseed Producers Cooperative in Michigan (non-GMO soybean products for human consumption), and the Organic Valley of Farms headquartered in Wisconsin (organic dairy products).[6] Bilateral linkages describe collaborative business forms in which a central enterprise assembles the food through contracts with a set of producers that usually guarantee above-market prices in exchange for specified production standards. Examples of such bilateral linkages in food value chains include Niman Ranch, featuring pork products, and Laura's Lean Beef.[7]

High Levels of Performance and Trust

Successful value chains distinguish themselves by high-quality, differentiated products or services and high levels of performance throughout the network. This is done by establishing continuous improvement systems, high levels of assistance for strategic suppliers, and performance evaluation systems that engage the entire chain as well as focus on the quality of the final product and customer satisfaction as the fundamental units of evaluation. Instructive examples come from the automobile industry, which in many ways has pioneered the value chain business model. Toyota, an industry leader, employs six different yet complementary strategies to heighten learning and improve performance throughout its network of suppliers: a strategic supplier association within which a great deal of learning and sharing of technical knowledge occurs; consulting teams of experts that regularly visit suppliers and are available on request; temporarily assembled teams to solve special problems in the system; interfirm employee transfers in which employees of Toyota spend periods of time working in supply firms and vice versa; voluntary study groups of suppliers operating in similar geographic areas; and finally, performance feedback and monitoring processes that engage the total supply chain (Dyer 2000, 63–85).

The kind of information sharing underlying these performance systems in value chains is possible only if high levels of trust undergird business relationships among the strategic partners. In fact, trust is pointed to as a pivotal component in successful value chains by virtually all observers of these interorganizational alliances (van Donkersgoed 2003, 1; Kumar 1996, 98; Schotzko 2000, 18; Zuurbier 1999, 24; Whalen 2001, 84, 121; Kaplinsky and Morris 2001, 74; Dyer 2000, chapter 4; Handfield and Nichols 2002, 163–174). While dimensions of interorganizational trust can be derived from studies of interpersonal trust (Handfield and Nichols 2002, 164), it is important that trust in value chains be based not on personal relationships but organizational procedures, or "process-base trust" (Dyer 2000, 180). In other words, it is a trust in the fairness, stability, and predictability of the procedures and agreements among strategic partners; and that policies are consistent and stable over time, and do not change with new management or personnel. A working definition of interorganizational trust is the mutual confidence of enterprise partners that each will fulfill its agreements and commitments, and will not exploit the other's vulnerabilities (ibid., 88). Key elements of organizational behavior that generate and maintain trust include reliability, fairness, competence, goodwill, loyalty, and respect for the risks and vulnerabilities associated with business models based on interdependence (Handfield and Nichols 2002, 163–173; Dyer 2000, 89; Kumar 1996, 96–98).

Finally, interorganizational trust takes time to develop, works best when it's mutual, is usually limited to specific areas of business, needs to be associated with consequences for breaking trust, and yet creates a reservoir of goodwill that helps preserve partnerships when, as will inevitably happen, one party engages in an act that its partner considers threatening (Kumar 1996, 95–96). In some advanced value chains, a powerful expression of trust is minority ownership positions among strategic partners. Toyota, for instance, owns roughly 30 percent of the shares of its major suppliers—an important practice for creating an attitude of mutual destiny and cooperation within the extended enterprise. As one analyst observes, "This is a significant enough ownership stake to build goal congruence—and trust—between Toyota and its supplier partners" (Dyer 2000, 107–108).

The implications of performance and trust issues for midtier food value chains are several. First, the reliable production, processing, and distribution of high-quality food products are essential. This can be problematic as the quality and consistency of many food products is dependent on weather, seasonality, grower's and/or processor's competence, and the availability of quality-preserving distribution mechanisms (Zuurbier 1999, 21). Quality assurance systems (with realistic allowances for surprise events) will need to be put in place throughout the value chain, with particular attention paid to the production and processing links. Such quality assurance systems are especially important because these are the sectors where food craftspersons like skilled farmers and specialty food processors can generate significant value through the application of skills that are complex, difficult to duplicate, and likely to generate competitive advantages that are sustainable.[8] To provide consistently high-quality food products, new farming or ranching and processing techniques will need to be developed and/or adapted for midtier value chains. Animal husbandry examples include new ultrasound techniques for predicting and measuring beef quality, new methods for predicting pork quality through measuring the pH levels in the meat, or the use of air-chilling methods that result in better-tasking poultry products, as is done in Europe.[9]

Second, as regionally based midtier food value chains begin to emerge throughout the country, developing the capacity to employ a number of the learning and performance-enhancing strategies associated with experienced extended enterprises, like Toyota, would seem to be in order. Support infrastructure such as a national network to assist midtier food value chains, for instance, could productively facilitate the sharing of knowledge between beginning and more experienced food alliances, establish consultant or "swat teams" to help developing value chains address problems, or sponsor regional workshops or study groups to upgrade the procedures of participating enterprises.

Finally, having struggled in arm's-length or adversarial market contexts for years, farmers and ranchers are understandably likely to be suspicious of business models based on trust and interdependence. Their participation in new midtier food value chains will be greatly fostered by an awareness of successful food chains, analyses of the lessons learned from unsuccessful ones, and the patient development of chained enterprises that share decision making as well as respect partners' risks and vulnerabilities.

Shared Vision, Information, and Decision Making

Given the high levels of trust and interdependence associated with successful value chains, it is critical that all partners in the extended enterprise share a common vision as to product quality, partner relationships, and customer treatment. Companies that want to develop strategic alliances pay great attention to partner selection (Handfield and Nichols 2002, 159–161).

Related to and ranking with trust as a key characteristic of successful value chains is effective information flow (Boehlje et al. 1999, 2; Dyer 2000, chapter 3; Handfield and Nichols 2002, chapter 9). A lack of information or distorted information passed from one end of the supply chain to the other can create significant problems, including but not limited to increased design time, misguided capacity plans, missed production schedules, excessive inventory investment, ineffective transportation, poor customer service, and lost revenues (Handfield and Nichols 2002, 295). On the other hand, effective information flows—called "information visibility" (ibid., 298)—enable value chain partners to accurately share forecasts, manage inventories, schedule work, and optimize deliveries. In doing so, the partners can reduce costs, improve productivity, and create greater value for the final customer in the chain (ibid., 258).

In addition, effective information sharing enables partners to evaluate performance, detect problems, and engage in problem solving (ibid., 258). Such capacities enable successful value chains to learn and respond faster than their competitors (Dyer 2000, 61–63). Advances in computer-based information technologies anchored by the World Wide Web and/or the Internet as well as dedicated communication networks within and between business partners (intranets and extranets) greatly facilitate the effective sharing of strategic information.[10]

Shared decision making and effective governance are also the marks of successful value chains (Boehlje et al. 1999, 2; Kaplinsky and Morris 2001, 29–32, 66–69). The issue of shared governance is where the distinction between suppliers as providers and suppliers as partners is joined. As one observer put it, "Real partnership means all participants benefit and all have a say in developments" (van Donkersgoed 2003, 1). This does not mean, however, that all partners will have equal power in strategic business alliances. In reality, most value chains are unbalanced with regard to power (Kumar 1996, 96; Kaplinsky and Morris 2001, 29). What is important for effective governance is that legitimated mechanisms are put in place for the following types of governance: legislative (setting standards for the supply chain), judicial (monitoring performance in the supply chain), and executive (coordinating procedures and flows in the supply chain) (Kaplinsky and Morris 2001, 31). These mechanisms can be located within the value chain or external to it, as is the case with third-party monitoring and certification of standards. Additionally, to be effective the power to govern means the capacity to sanction behavior within the value chain, either by imposing punishments or providing rewards (ibid., 30). Finally, managing value chains can require significant time, effort, and resources (Handfield and Nichols 2002, 177).

Power and authority issues within value chains are important. Studies of successful value chains indicate that governance can be effective in strategic business alliances when the less powerful partners experience the governing actions of the more powerful partner as fair or just (Kumar 1996, 96). Fairness encompasses two types of justice. The first is distributive justice, or the perceived fairness of the outcomes received. (This kind of justice will be dealt with in the following section of this chapter.) The second is procedural justice, or the perceived fairness of the powerful party's process for managing the relationship. Value chains that are experienced as procedurally just are built on the following kinds of principles (ibid., 97–98):

• Bilateral communication: The more powerful party is willing to engage in open and honest two-way communication with its partners
• Impartiality: The more powerful party deals with all its partners equitably
• Refutability: The less powerful or more vulnerable partners can appeal the more powerful partner's policies and decisions
• Explanation: The more powerful party provides its partners with a coherent rationale for its decisions and policies
• Familiarity: The more powerful party understands or is aware of the conditions under which its partners operate
• Courtesy: The more powerful party treats its partners with respect

Midtier food value chains will need to engage the issues of shared vision, shared information, and shared governance and decision making in ways that are appropriate for their respective alliances. An emphasis on building regional food economies is likely to be a crucial part of the vision and effectiveness of midtier food value chains. Regionality contrasts with and complements well the focus on the local for direct food marketing systems and on the global for commodity food marketing strategies. The Iowa-based Regional Food Systems Working Group has developed a working definition of regional food systems:

A regional food system supports long-term connections between farmers/ranchers and consumers while helping to meet the health, social, economic, and environmental needs of the communities within the region. Producers and markets are linked via efficient infrastructures that do the following:

- Promote environmental health
- Provide competitive advantages to producers, processors, and retailers
- Encourage identification with the region's culture, history, and ecology
- Share risks and rewards equitably among all the partners in the system.[11]

Regional and local food distribution systems offer valuable environmental benefits compared to national or global food distribution systems by using less fuel, because the distance from farm to consumer is shorter in the local and regional systems. In a 2001 study, a state-based regional food system used seventeen times less fuel than a national-based food distribution and transport system, and four times less fuel than a local system represented by farmers' markets and CSA enterprises (Pirog et al. 2001). Interestingly, the regional distribution system was significantly more fuel efficient than the local system.

In addition to fuel efficiencies, regionally operative midtier food value chains are likely to have advantages in strategic information flows as the ability to track and monitor deliveries is highly (and negatively) correlated with distance (Handfield and Nichols 2002, 316). Accurate and up-to-date (real-time) information is particularly critical for food products where weather and other factors can create considerable variation and uncertainty (Zuurbier 1999, 21; Schotzko et al. 2000, 19). Partnering with businesses that have regional decision-making authority should magnify communication advantages.[12]

While successful midtier food value chains will treat all strategic enterprises as partners, the distinction between suppliers and partners is particularly important for farmers or ranchers and midtier food processors. Both of these links in the value chain are accustomed to operating as cheap input suppliers in adversarial and/or arm's-length market relationships. The new food value chains will need to build business scenarios that provide farmers or ranchers and local food processors with some level of countervailing economic power as well as agreements and procedures to govern relationships with more powerful partners that are experienced as fair and just.[13]

This will be particularly true for start-up value chains that engage a large, established food company as a strategic partner. Reflecting on the historically adversarial paradigms of many agrifood supply chains, observers of the contemporary food system critically refer to these disproportionately influential partners as "channel masters" (Peterson 2002, 1333–1335). An important mechanism for farmer or rancher empowerment is their retention of control of the food product throughout the value chain, either through actual ownership or maintenance of a farmer- or rancher-based brand through to the consumer.

As is the case with food value chains based on organic production systems or fair trade, the use of standards and third-party certification will also be essential for new midtier food value chains. Organizations with experience in framing and certifying standards will play key roles in the governance structure of these new food chains. Particularly important and challenging will be establishing standards for all the partners in the value chain, and engaging such disparate dimensions as food quality, environmental stewardship, animal care, workplace conditions, and business ethics.

Support for Strategic Partners

Japanese firms with extensive experience in value chains talk about the underlying philosophy as "co-existence and co-prosperity" (Dyer 2000, 159). As indicated above, successful value chains are based on the trust that each partner is interested in the other's welfare and neither will seek to exploit the other's vulnerabilities (Kumar 1996, 94). Arguing for the importance of transforming many traditional supply chains into "integrated value systems," one book concludes, "The only solution that is stable over the long term is one in which every element in the supply chain, from raw material to end consumer, profits from the business. It is shortsighted for businesses to believe they can solve their . . . problems by punishing suppliers and customers" (Handfield and Nichols 2002, 6).

Critical components of seeing to the welfare of all in the network are conscious strategies to maintain product differentiation and value. Such strategies differ fundamentally from commodification strategies based on achieving the lowest costs of production globally—known as the "race to the bottom" or "immiserising growth" (van Donkersgoed 2003, 2; Kaplinsky and Morris 2001, 21–22). In addition to the continuous quality improvement of products, value-maintaining strategies also focus on ensuring fair economic rewards across the strategic network.

Value chain economics are based on three interrelated components. First, strategic partners are rewarded based on agreed-on formulas for adequate margins above production costs and adequate returns on investment. This differs markedly from low-cost bidding mechanisms and requires partners to have a good handle on their true cost structure. It also requires a high degree of information sharing regarding sensitive economic data—one of the most challenging dimensions in successful value chain creation (Handfield and Nichols 2002, 234). Such transparency with regard to economic information sharing accurately points to the importance of trust and procedural fairness within the network. Parallel to fair margins are fair and livable wages for workers employed by value chain partners, including fringe benefits and promotion opportunities. Emerging models of sustainable value chains also include community-level economic characteristics like opportunities for investment in food-processing enterprises by local community residents.[14]

The literature refers to the second component of value chain economies as "target- or cost-based" pricing (Dyer 2000, 121–122; Handfield and Nichols 2002, 236–238; Kaplinsky and Morris 2001, 62). The allowable costs of production along the value chain are determined by calculating backward from a selling price that it is estimated a customer will pay. Agreed-on profit margins are then subtracted from this selling price, with the allowable costs of production determined by what remains. This approach not only builds in the supplier's profit requirements up front but sets the stage for all partners to share information regarding ways to reduce costs based on an understanding that the benefits of such cost reductions will be shared across the chain. This cost-based approach provides incentives for partners to pursue continuous performance improvement to realize shared cost savings and invest in productive assets (Handfield and Nichols 2002, 236). As the CEO of a large automobile company noted in an address to supply partners, "All I want is your brain power, not your margins" (Dyer 2000, 124).

The third component of value chain economics involves contracts and agreements of an appropriate, extended duration. Successful value chains are built on long-term partnerships (Kumar 1996, 95). In addition to fair profit margins and a share of cost savings, strategic partners must be confident that their business is safe if they are to continuously improve performance and invest in more productive assets (Dyer 2000, chapter

2). In well-established value chains, new contracts with strategic partners are awarded on past performance, not on competitive bidding (ibid., 101–102; Kaplinsky and Morris 2001, 74). In the Toyota value chain, strategic suppliers are assured of contracts for the life of a vehicle model, as long as the quality of their work is high and they meet agreed-on target costs (Dyer 2000, 121–122). Once trust is established in value chains, legal contracts tend to be simplified, or in some cases, replaced by informal agreements. As one author put it, "What holds these relationships together is not legal force, but mutual obligations and opportunities" (Kumar 1996, 98).

Several midtier food value chains created in the last two decades provide strong examples of the economic components highlighted above. One is the Oregon Country Natural Beef cooperative.[15] Established in the late 1980s, this co-op of 120 ranching families in the Pacific Northwest engages approximately seventy-eight thousand cow-calf pairs (all marketed cattle raised from birth) and 3 million acres of rangeland. Grazed through most of their lives, the beef animals are antibiotic- and hormone-free, and finished for three months on a GMO-free and reduced-grain ration. In 2006, Oregon Country Natural Beef marketed the meat from nearly nine hundred animals per week. Partners in this value chain include a custom feedlot owned by an Oregon Country Natural Beef member rancher, a midsize meatpacker, and marketing partners that include regional supermarkets and a regional fast-food restaurant chain, enabling the cooperative to market all cuts of meat.[16]

Exhibiting the economic components of true value chains, prices paid to the ranchers are based on the "cost of production, return on investment, and a reasonable profit" to ensure the economic viability of member ranches. When conventional market prices for beef rose above Oregon Country Natural Beef's pricing levels in late 2003 due to the BSE case discovered in the state of Washington, the cooperative did not change its prices because the group felt that fluctuating prices had little long-term bearing on whether ranchers could stay on the land. Oregon Country Natural Beef also exhibits the second economic component— the continuous improvement of value chain efficiencies—by working with its strategic partners to streamline the production, feeding, and slaughter operations. Finally, the co-op conducted $45 million in retail business in 2006 without a single legal contract—affirming that Oregon Country Natural Beef is engaged in an effective, interdependent value chain characterized by high levels of trust and experienced mutual benefit among all strategic partners.

A second example is also from the Pacific Northwest: Yakima Chief Inc., a hop-producing, processing, warehousing, and marketing company.[17] Formed in 1997 and owned by thirteen hop-growing families, Yakima Chief provides specialty hop products to breweries worldwide, including microbreweries. Yakima Chief customers include partners who receive highly customized service and a guarantee of supply. The price of the hop products is determined by a formula that provides for pricing based on the average cost for an average level of productivity plus a fixed profit margin. Business relationships between the strategic partners are viewed as long-term and durable, with compelling reasons needed to undo business agreements. Continuous improvement systems are in place with reporting tools, and monthly grower meetings offer assistance to hop growers to improve the efficiency of their farming systems. Systems of traceability back to individual growers' fields supply assurance of quality and food safety. Information sharing occurs at annual meetings between team members from each partner, and via ongoing dialogue throughout the year with regard to sales and pricing issues. These mechanisms enable the partners to identify and respond to challenges quickly as well as effectively. Yakima Chief's model demonstrates that farmers who offer a value proposition that provides exceptional customized service and a guarantee of supply can, in most instances, be assured of obtaining a significant market share at premium prices. A grower in such a value chain may expect a reasonable level of profit above their fixed and variable production costs if their productivity is average or higher with a cost structure that is average or below.

Challenges and Recommendations for Midtier Food Value Chains

After five years of doing business, the Tallgrass Prairie Beef Cooperative —a rancher-based enterprise in Kansas—ceased operation for economic reasons in 2000. The cooperative's former business manager shared a series of challenges for midtier food value chains (Wilson 2001). This chapter concludes by examining these issues. The discussion focuses on how a set of midsize food enterprises, occupying various positions in food value chains and at various stages of development, are addressing the challenges that undermined the Tallgrass Prairie Beef Cooperative. These enterprises were selected as case studies by a national task force on regenerating an agriculture of the middle.[18]

Challenge 1: Differentiation and Pricing Strategies
The Tallgrass Prairie Beef Cooperative concentrated its product differentiation on offering grass-finished beef for sale. Not only were the environmental and human health benefits of such meat little known to consumers in the mid-1990s but grass finishing resulted in uneven grading among animals and increased on-ranch costs of production. Contrast this to the effective product differentiation of Oregon Country Natural Beef, which focuses on the antibiotic- and hormone-free qualities of its meat. While raised on pasture, Oregon Country Natural Beef animals are finished on a low-grain diet, which results in meat that grades consistently and fits well into the USDA beef-quality grading system—a context familiar to consumers. Successful midtier food enterprises have all discovered effective dimensions of product differentiation from the heritage animal genetics emphasized by the Ozark Mountain Pork cooperative in Missouri, to the premium-quality wines of New York State's winery associations or the human-food-grade soy products manufactured by the Thumb Oilseed Cooperative in Michigan, to the multiple ages and varieties of cheddar cheese (including kosher) sold by the Tillamook County Creamery Association in Oregon. While food character and quality are the primary axes of differentiation, regional production is an important marketing component for some new enterprises, including Farm Fresh Connections, which sells beef and vegetables to private colleges in Maine, and a group of grain farmers and millers in Washington who sell specialty flours to artisanal bakeries. Effective midtier food value chains will identify and explore differentiated products based on a dialogue between producers and partners who are in close touch with markets and customers.

Pricing strategies for midtier food value chains will seek to secure for farmers and ranchers prices that are above wholesale levels, but likely below those prices received from selling directly to consumers. This market fits midsize farms that want to shift to or expand their production of higher-value food products and need a premium over commodity markets. Premiums are achieved and maintained through multiple strategies including high product quality, brand recognition, and the control of production levels to match market demand, as is done successfully by experienced "new-generation" cooperatives like the Organic Valley of Farms, headquartered in Wisconsin. True value chains reward strategic partners with prices that are based on adequate margins above production costs and adequate returns on investment. As seen in the value propositions supporting both the Oregon Country Natural Beef and the Yakima Chief enterprises, all partners need to be well versed in their respective production cost structures, including honest accounting for time spent and the ability to use accrual accounting in business management.[19] This is likely to be an initial challenge for some farmers and ranchers aspiring to value chain participation.

Challenge 2: Achieving a Sufficient Volume and Supply of High-Quality Food Products

As the literature on value chains emphasizes and the more established midtier food enterprises demonstrate, sufficient scale is an important dimension for these business networks. For example, the Tillamook cooperative accounts for nearly one-third of the dairy products produced in Oregon, over 2.5 million people visit New York's regional wineries every year, and the Organic Valley Family of Farms engages over five hundred farming families and nearly twenty thousand dairy cows. Niman Ranch has expanded from a handful of Iowa pork producers in 1996 to nearly 450 pork producers in nine states. Efficiencies of scale are related to such matters as the securing of good rates for the transportation and slaughter of animals as well as the building of cost-effective assemblying and distribution systems for produce and dairy products. With too few animals to offer meat processors, the Tallgrass cooperative paid triple the processing rates of higher-volume suppliers. Achieving these efficiencies is particularly challenging in the earlier stages of enterprise development when volumes tend to be relatively low. For the above reasons, it will be critical for new midtier food value chains to piggyback whenever possible on the existing assemblying and distribution infrastructure of strategic partners, like food service companies. Finally, technical and financial assistance to farmers to enter or transition into new value chains is likely to be necessary in many situations to ensure sufficient quantities of differentiated food to meet the demand.

Related to volume is a consistent supply of high-quality food products moving through the chain. Quality control mechanisms will need to be operative throughout the strategic network, with particular attention paid to quality and consistency at the producer end, given the collaboration of multiple individual farms and ranches in the new midtier food value chain model. In addition, quality control and consistency will be challenges in situations where seasonality affects food production. Challenged to provide a year-round consistent supply of beef, the Oregon Country Natural Beef cooperative employs winter strategies like backgrounding animals on small grains and/or holding cattle to deliver to the feedlot at heavier than normal weights. Backup supply sources will also need to be in place, as evidenced by Yakima Chief's strategy of buying raw hops in the marketplace to meet contractual commitments for hop varieties outside the owner-grower's crop portfolio—for instance, to meet a microbrewery demand for a hop produced abroad in England, Germany, or the Czech Republic. Finally, experienced food value chains, like the successful "red label" poultry system in France, put in place traceability mechanisms that provide capacity to trace food products back through the chain, including to the site of production. These systems reassure consumers of the food's quality because a food taste or safety problem is rapidly traced, located, and solved (Stevenson and Born 2007; Westgren 1999, 1110).

Challenge 3: Adequate Capitalization and Competent Management

Securing adequate investment capital can pose significant challenges for beginning midtier food enterprises. As experienced by the Ozark Mountain Pork cooperative, the capital needs of midsize enterprises can be too large for smaller local banks, yet too small for larger national lending institutions, which offer the best interest rates. The Texas Organic Cotton Cooperative and Farm Fresh Connections in the Northeast have experienced parallel difficulties raising operating and expansion capital. For midtier food value chains to flourish, strategies will need to be developed to access capital, particularly bridge funding during the latter stages of the business start-up to self-sufficiency stage. Even more established mid-scale food enterprises can face challenges with raising capital, particularly cooperatives like the Organic Valley of Family Farms that have more legal restrictions on raising capital than competing private food companies.

Competent management is especially important for developing midtier food businesses because key dimensions of the value chain business model contrast with the conventional arm's-length or adversarial approaches to business. Decision makers for such enterprises as the New Seasons Market, Yakima Chief, and Oregon Country Natural Beef reveal solid understandings of such issues as interorganizational trust, shared information, fair prices, and long-term business relationships. Ozark Mountain Pork, on the other hand, has had difficulty finding managers who understand these new ways of doing business. That is one of the reasons that Oregon Country Natural Beef has elected thus far to hire its management team from among its rancher members.

Related to management issues are questions about the organizational models that most fully support the respective competencies of farmers and ranchers as well as managers in other enterprises in the value chain. More specifically, models are emerging in new food value chains wherein producer cooperatives focus on farming or ranching yet own minority stock positions in downstream partner enterprises. As indicated earlier in this chapter, cross-ownership among value chain partners provides several collaborative advantages and is not uncommon in well-established value chains. The clearest case of such coupling in new food value chains is the ownership position taken by the Texas Organic Cotton Cooperative in its partner, Organic Essentials, a private company that makes and sells organic personal care products such as tampons, swabs, cotton balls, and nursing pads.

Challenge 4: Technical, Research, and Development Support

Land grant universities have traditionally supported faculty and extension service positions to help develop and promote the use of commodities, such as corn and soybeans in the upper Midwest as well as wheat in the Great Plains. Currently, these universities, beset by state budget cuts and the elimination of positions, are struggling to transition their faculty and staff to respond effectively to the challenges inherent in a diverse set of highly differentiated food value chains. The sustainable agriculture and value-added programs at these universities also for the most part have failed to engage their colleges of business in addressing these challenges. Yet models are being developed within parts of the land grant system to link agrifood and business professionals to address the challenges and opportunities found in these new midtier food value chains.[20]

Challenge 5: Creating Meaningful Standards and Consistent Certification
Many well-established value chains have developed standards for performance at each organizational link in the chain, and employ neutral third-party agents to certify the adherence to these standards. For example, the red label value chain system for high-quality poultry products in France requires a minimum of sixty-five tests along the supply chain, from farmers to retailers (Stevenson and Born 2007). Such standards and certification procedures will need to be developed for midtier food value chains in the United States. We conclude this chapter by sharing a proposed value chain pledge of interdependence and performance that clearly identifies key dimensions for standards setting.

Agriculture of the Middle: A Value Chain Pledge of Interdependence and Performance

Vision Statement
Our goal is to encourage the creation of economic value chains distinguished by a mutual commitment to sustainability, fairness, and food quality. All partners in the value chain pledge to make business decisions that will ensure the economic sustainability of all other partners in the chain.

The success of values-based business chains will be measured by increases in the volume of the food sold by companies that are committed to food-quality enhancement, environmental and resource stewardship, transparency, and the equitable sharing of power and economic returns across the value chain.

A Values-Based Business Chain Pledge

Farmer Pledge (*Name of farm*) pledges to work as a partner within value chains dedicated to the sustainable production of high-quality, differentiated food products along with the delivery of fair prices and equitable financial returns across the value chain. (*Farm name*) commits to design as well as manage farming and animal care systems that conserve natural resources and environmental quality, promote animal health, and sustain healthy farms, workplaces, and rural communities. Sharing production and marketing information with other partners in the value chain, (*farm name*) will help develop and assure compliance with mutually agreed-on standards of food quality, production systems, animal care, workplace conditions, and business ethics, based on third-party certification. Such standards and ethics will help define the unique quality attributes associated with food moving through the value chain, with the goals of customer satisfaction as well as capturing and sharing added economic value in the marketplace.

Food Processor Pledge (*Name of company*) pledges to work as a partner within value chains that are dedicated to the production of high-quality, differentiated food products along with the delivery of fair prices and equitable financial returns across the value chain. Sharing production and marketing information with other partners in the value chain, (*company name*) will help develop and assure compliance with mutually agreed-on standards of food and environmental quality, production systems, animal care, workplace conditions, and business ethics, based on third-party certification. Such standards and ethics will help define the unique quality attributes associated with food moving through the value chain. Verifiable quality attributes will be featured in product labeling, marketing, and consumer education campaigns, with the goals of customer satisfaction as well as capturing and sharing added value in the marketplace.

Food Distributor Pledge (*Name of company*) pledges to work as a partner within value chains that are dedicated to producing, distributing, and marketing high-quality, differentiated food products along with the delivery of fair prices and equitable financial returns across the value chain. Sharing distribution and marketing information with other partners in the value chain, (*company name*) will help develop and assure compliance with mutually agreed-on standards of food and environmental quality, production systems, workplace conditions, and business ethics, based on third-party certification. Such standards and ethics will help define the unique quality attributes associated with food moving through the value chain. Verifiable quality attributes will be featured in product certification and labeling, marketing, and consumer education campaigns, with the goals of customer satisfaction as well as capturing and sharing added value in the marketplace.

Food Retailer Pledge   (*Name of company*) pledges to work as a partner within value chains that are dedicated to the production and marketing of high-quality, differentiated food products along with the delivery of fair prices and equitable financial returns across the value chain. Sharing marketing information with other partners in the value chain, (*company name*) will help develop and assure compliance with mutually agreed-on standards of food and environmental quality, production systems, animal care, workplace conditions, and business ethics, based on third-party certification. Such standards and ethics will help define · the unique quality attributes associated with food moving through the value chain. Verifiable quality attributes will be featured in product certification and labeling, marketing, and consumer education campaigns, with the goals of customer satisfaction as well as capturing and sharing added value in the marketplace.

(This pledge is intended to include other agriculturally based goods such as fiber- or energy-related products.)

Notes

1. See the results of the consumer surveys available at ⟨http://www.leopold.iastate .edu/pubs/staff/files/050504_ecolabels2.pdf⟩;   ⟨http://www.organicvalley.coop/ mediacenter/press_release_detail.php?id+210⟩; ⟨http://www.farmprofitability.org/ local.htm⟩.

2. In addition to the business organizational literature, which focuses significantly on strategic alliances in the automobile industry, our discussion of value chains is also informed by the fair-trade literature (Jaffee et al. 2004), and the idea of "associated economics" employed by followers of the early twentieth-century political economist Rudolph Steiner (Lamb 1996, 1997; Perlas 2003; Steiner 1985).

3. See ⟨http://www.valuechains.org/valuechain.html⟩.

4. New Seasons Market is featured along with other case studies of new midtier food enterprises on the Web site of a national task force on renewing an agriculture of the middle: ⟨http://www.agofthemiddle.org⟩.

5. Examples of strategic components in the automobile industry that differentiate car models include transmission and engine parts, body and instrument panels, and seats. Nonstrategic parts include rubber hoses, belts, and tires (Dyer 2000, 152).

6. These four food cooperatives are included among the case studies featured on the national task force Web site: ⟨http://www.agofthemiddle.org⟩.

7. See ⟨http://www.nimanranch.com⟩; ⟨http://www.laurasleanbeef.com⟩.

8. Author Jeffrey Dyer calls these craft skills "tacit know-how" (2000, 63–64).

9. For new ultrasound techniques being developed to evaluate beef quality, contact the Jacob Alliance (⟨beefrepro@aol.com⟩). For pH measurements as predictors of pork quality, contact Ken Prusa, Iowa State University Food Science and Human Nutrition Department (⟨kprusa@iastate.edu⟩) For reference to the air-chilling processing systems used throughout Europe that result in better-tasting poultry meat than do the water-chilled systems primarily employed in the United States, see Stevenson and Born (2007).

10. For a description of the sophisticated information system that Dell computer company employs to communicate with its customers and business partners, see Handfield and Nichols (2002, 259).

11. See ⟨http://www.valuechains.org/rfswgf/rfs_definition.html⟩.

12. Examples of national food companies that have regional decision-making structures include Organic Valley (⟨http://www.organicvalley.com⟩) and SYSCO (⟨http://www.sysco.com⟩).

13. For a description of eight major areas that should be addressed in contracts between farmers or ranchers and processors or integrators, see Moline (2004).

14. See the draft matrix of sustainability characteristics for bio-based value chains (⟨rspirog@iastate.edu⟩).

15. See the description of Oregon Country Natural Beef available at ⟨http://www.agofthemiddle.org⟩.

16. For information on two of Oregon Country Natural Beef's value chain partners, New Seasons Market and Burgerville, see ⟨http://www.newseasonsmarket.com⟩ and ⟨http://www.burgerville.com⟩, respectively.

17. See ⟨http://www.yakimachief.com⟩.

18. See ⟨http://www.agofthemiddle.org⟩.

19. Accrual accounting is based on recording income as accounts receivable when earned and recording debts as accounts payable when they are incurred.

20. Examples of new land grant university-based developments include the Value Chain Partnerships for a Sustainable Agriculture project led by the Leopold Center, Practical Farmers of Iowa, and Iowa State University as well as a project focusing on the production and marketing of specialty cheese products involving

the Center for Integrated Agricultural Systems at the University of Wisconsin at Madison and the Dairy Business Innovation Center located in the Wisconsin Department of Agriculture, Trade, and Consumer Protection (see ⟨http:// www.valuechains.org⟩; ⟨http://www.cias.wisc.edu/marketing.php⟩; ⟨http://www .datcp.state.wi.us/mktg/business/marketing/val-add/initiative/pdgf⟩).

References

Boehlje, Michael, Steven Hofing, and R. Christopher Schroeder. 1999. Value Chains in the Agricultural Industries. Staff paper no. 99–10. West Lafayette, IN: Department of Agricultural Economics, Purdue University.

Callimich, Rukini. 2004. Caterer Turns the Tables. Available at ⟨http://www .registerguard.com/news/2004/08/10/b1.bz.bonappetit.0810.html⟩.

Dyer, Jeffrey. 2000. Collaborative Advantage: *Winning through Extended Enterprise Supplier Networks*. New York: Oxford University Press.

Fearne, Andrew, David Hughes, and Rachel Duffy. 2001. Concepts of Collaboration in Supply Chain Management. In *Food Supply Chain Management*, ed. Jane Eastham, Liz Sharples, and Stephen Ball, 55–89. London: Butterworth and Heinemann.

Greenberg, Laurie. 2004. Working with Retail Buyers. Report published by the Center for Integrated Agricultural Systems, University of Wisconsin at Madison. Available at ⟨http://www.cias.wisc.edu⟩.

Handfield, Robert, and Ernest Nichols Jr. 2002. *Supply Chain Redesign: Transforming Supply Chains into Integrated Value Systems*. Upper Saddle River, NJ: Prentice Hall.

Jaffee, Daniel, Jack R. Kloppenburg Jr., and Mario B. Monroy. 2004. Bringing the "Moral Charge" Home: Fair Trade within the North and within the South. *Rural Sociology* 69 (2): 169–197.

Kaplinsky, Raphael, and M. Morris. 2001. *A Handbook for Value Chain Research*. Available at ⟨http://www.ids.ac.uk/ids/global/pdfs/vchov01.pdf⟩.

Kirschenmann, Frederick. 2004. Are We about to Lose the Agriculture of the Middle? *Julien's Journal* (July).

Kumar, Nirmalya. 1996. The Power of Trust in Manufacturer-Retail Relationships. *Harvard Business Review* 74 (6): 92–107.

Lamb, Gary. 1996. Going beyond Self-interest in Economic Life. Part 1. *Threefold Review* (Winter–Spring).

Lamb, Gary. 1997. Going beyond Self-interest in Economic Life. Part 2. *Threefold Review* (Summer–Fall).

Moline, Stephen. 2004. Production Contracting: What a Producer Needs to Be Sustainable. Farm Division, Iowa Attorney General's Office, Des Moines.

Perlas, Nicanor. 2003. *Shaping Globalization: Civil Society, Cultural Power, and Threefolding*. Gabriola Island, BC: New Society Publishers.

Peterson, H. Christopher. 2002. The "Learning" Supply Chain: Pipeline or Pipedream? *American Journal of Agricultural Economics* 84 (5): 1329–1336.

Pirog, Rich, Timothy Van Pelt, Kamyar Enshayan, and Ellen Cook. 2001. Food, Fuel, and Freeways: An Iowa Perspective on How Far Food Travels, Fuel Usage, and Greenhouse Gas Emissions. Ames, IA: Leopold Center for Sustainable Agriculture.

Raynolds, Laura. 2000. Re-embedded Global Agriculture: The International Organic and Fair Trade Movements. *Agriculture and Human Values* 17:297–309.

Schettler, Ted. 2004. Nutrition and Food Production Systems: A Role for Health Care. Available at 〈http://www.sehn.org/ecomedessays.html〉.

Schnieders, Richard. 2004. Presentation at the Georgetown University Law School. Available at 〈http://www.agofthemiddle.org〉.

Schotzko, R. Thomas, and Roger A. Hinson. 2000. Supply Chain Management in Perishables: A Produce Application. *Journal of Food Distribution Research* (July).

Steiner, Rudolph. 1985. *The Renewal of the Social Organism*. Spring Valley, NY: Anthroposophic Press.

Stevenson, G. W., and Holly Born. 2007. The "Red Label" Poultry System in France: Lessons for Renewing an Agriculture-of-the-Middle in the U.S. In *Remaking the North American Food System*, ed. C. Clare Hinrichs and Thomas A. Lyson. Lincoln: University of Nebraska Press.

van Donkersgoed, Elbert. 2003. Value Chains versus Supply Chains. June 23. Available at 〈http://www.christianfarmers.org/commentary/cornerpost.htm〉.

Westgren, Randall. 1999. Delivering Food Safety, Food Quality, and Sustainable Production Practices: The Label Rouge Poultry System in France. *American Journal of Economics* 81:1107–1111.

Whalen, Charles. 2001. *The Featherbone Principle: A Declaration of Interdependence*. Gainsville, FL: Matthews Printing Company.

Wilson, Annie. 2001. Romance vs. Reality: Hard Lessons Learned in a Grassfed Beef Marketing Cooperative. *Rural Papers* 182 (October). Whiting: Kansas Rural Center.

Zuurbier, P. J. 1999. Supply Chain Management in the Fresh Produce Industry: A Mile to Go? *Journal of Food Distribution Research* (March).

IV

Policies That Could Support the Agriculture of the Middle

8

Toward a Pro-Middle Farm Policy: What Will It Take to Ensure a Promising Future for Family Farming?

Daryll E. Ray and Harwood D. Schaffer

Politicians perennially cite "saving the family farm" as a persuasive reason to renew income-supporting farm legislation. But as the dwindling number of farms would suggest, this country's farm programs may have solved other problems and met other pressing needs, but they have not stopped the loss of the moderate-size farms that many think of as family farms. They have not eliminated the need for producers on small farms to earn most of their household income from off-farm sources. What these programs have done is to provide price and income stability to compensate for market characteristics that cause chronic problems for the crop sector of U.S. agriculture. As we have seen, that alone does not ensure the viability and survivability of small and midsize farms. One purpose of this chapter is to identify private and public policies as well as strategies that are expressly designed to strengthen farming as a primary source of income for farms with less than $250,000 in annual sales.

After addressing definitional issues relating the term family farm, we begin by focusing on private and public policies that could help to undergird midsize farms or farms in the middle. Small farms, many of which sell directly to final customers or retailers, comprise another important segment of agriculture, especially in terms of the number of farms. After discussing private and public policies for this group of farms, we look at agriculture as an industry and discuss the challenges that large-scale agriculture will continue to face—specifically overcapacity and the sluggish response of the total agricultural quantities demanded and supplied to changes in prices.

The industry-level or more macro policies for agriculture address these long-known characteristics of aggregate agriculture, but these industry-level policies also form the foundation for building specific policies for small and midsize farms. It is important to understand that policies for targeted sizes of farms are unlikely to succeed separate from a well-designed set of macro agricultural policies.

Family Farm: Concept versus Definition

Directing policy specifically to midsize family farms requires a consensus on what is meant by midsize and what should be considered a family farm. As mentioned, framing farm legislation as a means of saving the family farm resonates well with the U.S. public. Even putting aside whether farm legislation has historically served that goal, the concept and definition of a family farm seems to be in the eye of the beholder. Part of the problem of defining a family farm is the fact that the conditions of agricultural production vary widely from crop to crop and region to region. Likewise, technological innovation has radically changed the labor and capital requirements of farming over the last three-quarters of a century. Following the Civil War, family farming for the newly freed slaves meant forty acres and a mule (Shannon 1989, 84). There was a time when family farming meant a farmstead on every quarter section of ground in many parts of the Midwest. Today a husband and wife team, using the latest horsepower and hydraulics, can manage a two-thousand-acre grain and hay operation in Kansas with little hired help.

Those seeking to lend some measurable substance to the term family farm have suggested three criteria: ownership, management, and labor. While the definitions vary from person to person, on a family farm the producer is generally expected to own at least some land, provide a majority of the other capital, and supply a majority of the labor and decision making (management).

Recently, Linda Lobao and Katherine Meyer used the term "farming as a household livelihood strategy" (2001, 104). When combined with ownership, management, and labor, the concept of farming as a household livelihood strategy provides a clearer picture of what many consider to be family farms. Currently, 95 percent of all farm household income comes from off-farm sources (Mishra et al. 2002, 16). Yet the benefits of current agricultural policies are skewed toward producers whose annual sales exceed $250,000.

In the past, the widely varying size and structure of agricultural operations has militated against finding a single set of agricultural programs
that meets the needs of all producers. To overcome that problem, this
chapter uses categories or groups of agricultural operations as focal
points for the formulation of agricultural policies that are tailored to specific agricultural situations and needs. The need to identify governmental
policies that support a structure of agriculture in which a family can engage in agriculture with the purpose of earning a livelihood from that
activity is the overall focus of this chapter. Since the policies envisioned
would be targeted to midsize and smaller farms, farm program benefits
would reach a more diverse set of agricultural producers than current
legislation. The next section focuses on midsize farms or farms in the
middle, and the following major section discusses the needs of smaller
farms that often are involved in direct selling.

Analysts have noted that there has been a hollowing out of U.S. agriculture. A few—but growing—number of large operations (with gross
sales above $250,000) produce most of the agricultural output, while
the farms that make up the biggest share of the total number of farms
are relatively small (under $100,000 in gross sales). It is the midsize
farms that are disappearing. These disappearing farms are likely to be
very ones that are operated by farm families who consider farming to be
a household livelihood strategy (Lobao and Meyer 2001, 106).

The challenge is to identify the characteristics of this group of midsize farms, and then a set of policies that utilize to advantage the skills
and resources of those farming operations. It is also important to consider the characteristics and policy needs of the large number of small
operations. In subsequent sections of this chapter, we will explore civic
agriculture and discuss some macro agricultural policy needs that span
all three farm sizes: midsize farms, civic agriculture, and large farm
operations.

Distinguishing Characteristics of Midsize Farms

The midsize family operation offers advantages in management skills,
flexibility, and adaptability. As we noted earlier, most observers have
considered one of the key characteristics of a family farm to be the
supplying of the majority, if not all, of the management for the farm

operation. In recent years, with contracting and vertical integration taking over the poultry and hog industries, and the GMO contracts that crop farmers have to sign to obtain access to the seed, the main management functions have been removed from the farmstead and placed in a far-off corporate office. With time, we expect to see these trends increase in operations of this sort. As contracting replaces the auction market in tobacco (Tiller 2001), for instance, we would not be surprised to see the management function effectively move to the tobacco company's corporate offices. Policies and strategies to strengthen farming of the middle need to be ones that use to advantage the management skills, flexibility, and adaptability of the individual operator.

Example or Model Arrangements

Recently, we have observed the movement of some operators away from low-cost and/or low-profit commodity production, and into tailoring their production to meet the needs of a well-defined specialty market. It is precisely this type of market that allows the midsize producer to use their skills to advantage. The market is too small for the large operators to bother with, yet large enough to provide a consistent demand for a group of producers who might work together in meeting its needs. For instance, some African American farmers in Mississippi are going into goat production to meet the needs of a growing Islamic immigrant population that prefers goat meat and desire to have it ritually slaughtered (Dagher 2004). To meet the needs of this market requires a degree of cooperation among farmers because it is larger than any one farmer can fulfill and yet too small for the integrators to find profitable. As long as the market size for this particular product remains in this intermediate range, it offers an opportunity for some producers to engage in agriculture as a household livelihood strategy.

In a similar effort, a group of cattle producers is organizing to provide beef for hospitals that want to serve meat that has not been raised with the prophylactic use of antibiotics. Small operations where the producer is actively involved in providing both the labor and management are in a much better position than large feedlots with tens of thousands of animals to manage the incidence of disease by identifying, isolating from the herd, and treating those cattle with veterinary problems. The challenge is to find the means of organizing a sufficient number of operators who will raise their cattle according to the needs of the end user—in this case a group of hospitals—by maximizing the management skills of the individual operators working in concert with each other.

Laura's Lean Beef is one example of how this kind of marketing can work to the benefit of midsize farmers. The following from the Laura's Lean Beef Web site gives some insight as to how this all works:

"Although the company has grown larger and more sophisticated, its priority is to remain true to its original values," [Laura] Freeman said. The family farm is at the heart of its operation. "We realize that it's more expensive for farmers to produce cattle to our specifications, so we pay a premium over market price," she said. "Quality, not quantity, is the key to economic survival for America's family farms."

Although the company has undergone eight logo changes due to brand development, the heart of its marketing effort remains direct communication with its customers. "We started our mailing list in 1985. Today it contains over 250,000 names," explains Freeman. The company's customer service representatives communicate with over 3,500 consumers each month: "Good communication between the people who produce food and their customers is part of America's farming tradition we think should be preserved," adds Freeman (2005).

Another model is the Organic Valley Family of Farms, which began in 1988 as a small, organic cooperative in Wisconsin. Today, Organic Valley consists of 619 organic farms in eighteen states from California to Maine. It markets products from milk to meat to vegetables, and has organized and marketed its products in such a way as to enable the producers to reap a greater portion of the retail dollar than general commodity production would. Its Web site strives to make connections with the consumers, and is replete with pictures of husbands and wives in farm settings, children with calves, and detailed descriptions of their agricultural practices. All of this combines to reestablish a partnership between the producer and consumer that has been lost as the commodity chains have become longer and longer. Organic Valley (2005) describes itself as a model for agricultural production and marketing.

One of the areas in which U.S. farmers are beginning to catch up with their European counterparts is the development and production of foods tied to a particular geographic location, a craft-based and more labor-intensive production process, or historical traditions. In Europe these foods include items like Modena balsamic vinegar, Parma hams, and various French wines. Although less common in the United States, the use of geographic indicators is growing with products like Vidalia onions and Florida oranges. The Leopold Center for Sustainable Agriculture at Iowa State University is promoting this idea among Iowa farmers who are looking for an alternative to commodity agriculture (Pirog 2004). For instance, Iowa's Maytag Dairy Farms (2005) offers a Maytag blue cheese produced with a craft-based, labor-intensive production process that allows it to maintain the distinctiveness of its product and command a premium price.

The production of identity-preserved crops also offers medium-size producers the opportunity to earn a premium over that available from growing a commodity crop like no. 2 yellow corn. In one sense, this is a return to a form of grain marketing that was common before the development of the Chicago Board of Trade in which each farmer's produce was bagged. The identity of the producer, the quality of production, and the location of the cropland was maintained all the way from the harvest to the final buyer. In that era, the crop was not commodified and the price received was, in part, related to the quality of the grain in the sack, since the produce's identity had been preserved (Cronon 1991, chapter 3). In today's marketplace, buyers are willing to pay a premium for quality production in which the identity is preserved from planting to harvest to final use. This requires extra care on the part of the producer, and represents an opportunity for those producers willing to provide that additional level of service.

This listing of strategies for midsize producers is intended to be suggestive, not exhaustive. It is meant to illustrate some key principles by which midsize producers can organize their farming operation to use to best advantage the characteristics that distinguish them from those who are producing for the undifferentiated commodity markets, allowing them to increase their share of the consumer dollar. One of the characteristics that those seeking to engage in farming as a household livelihood strategy have to exhibit is a keen understanding of the needs and/or desires of the consumer, whether that be concerns over the use of pesticides or GMOs, the need for a specialty product, the way the meat they eat is raised, the integrity of the people who are raising their food, or any number of individual expectations. It is the ability to respond to the needs of a particular set of customers that allows the midsize producer to offer a product and service that cannot be duplicated by their larger counterparts.

The midsize producer needs to be constantly attuned to the shifting desires and realities of the marketplace. As soon as the market share of any given specialty product becomes large enough, the large operators will figure out a way to break down the management skills of the midsize producer into a number of discrete tasks that anyone can perform and then commodify the production process. At that point the premium will disappear, and the midsize producer will need to move on to meet newly emerging consumer demands. We are seeing this today, with the commodity supply chains importing organic produce from Peru, for example.

To the extent that midsize farmers can identify a product that can be confined to a limited number of producers or a limited geographic area, the midsize producer may be able to avoid losing their markets to the large operators who are an integral part of the increasingly powerful supply chains (Hendrickson and Heffernan 2004). The key is to make sure that excess production does not drive down the price realized by the producer. The more that these producers can differentiate their product from the standard one, the more likely it is that they will be successful.

In summary, those engaged in farming as a livelihood strategy need to keep attuned to the following:

- Identifying and meeting consumer needs
- Segmenting the marketplace so as to distinguish their product from the commodity product
- Innovating in order to keep an edge over other producers
- Maximizing management skills
- Cooperating with similar-size producers to meet the needs of larger markets than can be met by an individual producer
- Controlling production to keep from commodifying their product

Policy Possibilities

The experiences of groups like Organic Valley, Laura's Lean Beef, and the numerous new-generation cooperatives that have developed in response to the increasing demand for ethanol should be looked at to see if changes in public policies and/or existing laws are needed to provide additional structures and/or more organizational flexibility for farmers seeking to work in concert with one another.

Extension programs could be built around doing feasibility and logistical studies; developing clearinghouses for producers, market participants, lawyers, accountants, and other professionals; providing educational programs on approaches to ensure consistently high-quality products that meet the expectations of the identified customers; and offering other facilitating services.

Federally sanctioned entities could be developed to handle some of these tasks, especially if legal protection was needed to accomplish the required collaboration among producers and other participants. Federal Marketing Boards supply a precedent for such federal structures, although the responsibilities and activities would likely be much different.

Certainly, publicly funded research could be called on to provide support to farmers who wish to identify and meet the needs of smaller markets, reestablishing the historic connection between producer and customer. Any technological advances so funded should remain in the public domain, not sold by private companies under the protection of patent laws. Other established government programs for agriculture, such as ensuring the ready availability of low-interest credit, could be shaped to specifically help midsize family farms.

Also, the federal government could be called on to enforce existing laws, and perhaps create new ones in the areas of market concentration, the environment, and labor-related issues, especially those that would primarily apply to larger operations. There is a continuum of possibilities in each of those areas. For example, in the environmental area, the list of possible activities could range from full enforcement of existing laws to requiring newly constructed confined-feeding operations to have waste disposal systems that meet the same standards as municipal waste treatment operations handling a similar volume.

Willard Cochrane (2003, 104) has suggested that "midsized-agriculture-in-the-middle" family farms that gross between $50,000 and $300,000 yearly should receive a no-strings-attached annual payment of $20,000. This payment would be a tangible expression of society's desire to preserve individual family farms and family farms in general. William Loughmiller (2005) provides a variation of Cochrane's plan. He suggests giving producers a $0.20 payment for every $1 of gross revenue from $30,000 to $230,000, which works out to a maximum payment of $20,000.

Civic Agriculture

"civic agriculture"

In addition to farms of the middle and large-scale commodity agriculture, which we will discuss later, there is an important component of U.S. agriculture that has been identified as "civic agriculture." In a recent article in *Rural Sociology*, Thomas A. Lyson and Amy Guptill (2004) contrast civic agriculture with commodity agriculture. While commodity agriculture is focused on providing an unending stream of an undifferentiated, standardized commodity to supply chains that reach around the globe, civic agriculture is a locally based agricultural production system focused on meeting the food needs of a relatively small area. It often uses direct sales to distribute its products. This segment has a different set of characteristics and policy needs.

"The organizational manifestation of civic agriculture include farmers' markets, community gardens, and community supported agriculture (CSA) and other forms of direct marketing" (Lyson and Guptill 2004, 371). Typically, civic agriculture is composed of small- to medium-scale farmers who are not able to earn a livelihood in extensive commodity agriculture. Rather than seeking to earn a small amount of money from each acre of a large operation, those engaged in civic agriculture farm the land intensively, concentrating on high-value production. Most of the farms engaged in civic agriculture have annual gross sales of less than $10,000 (ibid., 371).

What we are seeing in civic agriculture is the reintroduction of a form of production that gave New Jersey its nickname, "The Garden State." In the past, truck farmers, working on small family-size plots in New Jersey, provided New York City and Philadelphia with much of their agricultural produce. Today, CSAs around various population centers are growing in terms of both the quantity of food produced and the number of farmers who are turning to civic agriculture as a means of engaging in agriculture as a household livelihood option.

The needs of civic agriculture have not been a major concern of the triad of experiment stations, land grant colleges, and agricultural extension services that have been so much a part of commodity agriculture (Ostrom and Jackson-Smith 2005). Directing some of the funds of these agencies toward civic agriculture could pay rich dividends both in terms of the availability of sustainably produced local agricultural products and the opportunity for more small to midsize operators to earn a livelihood on their acreage. The August 2004 issue of Glynwood Center's (2005) *Gleanings* identifies a set of needs for farmers engaged in civic agriculture:

• "Access to new markets such as local restaurants, retail stores and institutional buyers, where the farmer can receive a fair price for his or her product"
• "An efficient distribution network that doesn't require the farmer to make the deliveries"
• "More local facilities such as community kitchens and slaughterhouses where farmers can produce value-added products"
• "Smarter consumers who understand the value of local food and appreciate that price is only one consideration"
• "Educated politicians and boards who understand how their policies and decisions either support or undermine farming."

Agriculture in Total: Issues and Policy Needs

Even though the central interest of many readers centers on farms in the middle and civic agriculture, it is critical to understand the issues and policy possibilities for agriculture as a whole, because the overall agricultural economic environment can make the difference between success or failure for operators of small and midsize farms. In many cases, the premiums negotiated by small and midsize family farm operators use the "commodity" version of their products as the beginning point, and therefore overall price levels in large-scale agriculture can be extremely important.

But what is it about agriculture that makes it different from other industries? Unlike automobiles, books, and computers, but like water and air, food is an absolute requirement for life itself. As a result, most governments exhibit an interest in food production that they show for few other products. While in the midst of World War II the U.S. government could convert automobile manufacturing lines to the production of armaments—leaving the public to find other means of transportation—the availability of food was ensured through the use of ration coupons.

The agricultural sector, particularly crop agriculture, is distinct from most other economic sectors in a number of crucial ways. The price elasticities of supply and demand are not sufficient to bring about a timely self-correction of the market (Ray et al. 2003, 21). Just as a diabetic does not purchase more insulin in response to a price decline, so most people do not increase their aggregate food intake from three meals a day to four in response to lower prices. A decline in the price of lumber may stimulate more do-it-yourselfers to take on the weekend project of building a new deck, but lower prices do not significantly increase the aggregate demand for food. Lower prices may stimulate people to eat out more often and pay for additional processing of the foods they prepare at home, but they do not significantly increase total food consumption (Ray 2001b).

Similarly, farmers tend to plant all of their acres under a wide range of prices. They may change the mix of crops in an attempt to maximize the revenue per acre, but they almost always plant all of their crop acreage particularly as long as the revenue per acre is above the out-of-pocket variable cost of production. Any dollar earned above that level can be applied to fixed costs like taxes. And on rented ground, the producer has every incentive to use every acre possible. It makes no sense to pay the cost of renting ground if the intention is to leave it unplanted. Unlike many other sectors where a few firms determine the size of the industry and can reduce production in an attempt to restore profitability, agriculture is composed of a large number of independent operations, none of which can affect either the price or industrial capacity (Ray 2001c). As a result, crop agriculture tends to use all of its productive capacity all of the time and let the weather determine the final production numbers.

One of the little-recognized factors in low crop prices is the role of public investment in research and extension in increasing supply at a faster rate than population growth. The planting of higher-yielding public varieties by producers is part of this expansion in supply, as is the adoption of management practices that have been identified through publicly funded research and disseminated through the work of the Cooperative Extension Service. The inevitable result of this supply increase in the face of an inelastic demand is lower prices. In this chapter, we are not at odds with the policy of public research in food production as a means of ensuring an abundant food supply for everyone. In fact, it would be immoral not to look for ways to ensure a sustainable supply of food adequate to meet the needs of the populace. If the government is going to interfere in the marketplace to increase the supply of food, however, then we would argue that it is appropriate for the government to put mechanisms in place by which that excess productive capacity can be managed for the long-run benefit of both producers and consumers.

From the earliest colonial period in the territory that became the United States through the 1920s, the primary public agricultural policies can be described as developmental. These policies were oriented toward the opening up and development of the agricultural lands of the country, and included land surveys, land sales, land grants to war veterans, land grants to companies to encourage the development of railroads to open up vast agricultural areas, and the granting of homesteads for individuals (Halcrow et al. 1994, 7). Developmental policies today continue in various forms, including farm credit programs, Rural Electric cooperatives' support for land grant colleges, and the funding of agricultural research and extension.

The 1930s saw the introduction of compensatory policies that provided price and income support for farmers. Initially the emphasis was on various mechanisms to support the price of selected commodities, indirectly providing support for producers. Typical of compensatory policies were ones that included programs to store surplus commodities during periods when production was greater than demand, programs to provide nonrecourse loans to farmers and thus establish a price floor, and acreage control programs to manage the use of the productive capacity of U.S. agriculture. In recent years, the stress has shifted to income-supporting programs that are decoupled from production (Ray 2001a).

The point of all this is to argue that agriculture is different, and the public policies a society chooses to put in place for crop agriculture need to be different from those one might use for restaurants, software developers, or pharmaceutical firms. For instance, the challenge for pharmaceutical firms is the high cost of developing new drugs and successfully getting them through the regulatory process. Some form of patent protection is therefore necessary if society wants the firms to continue to develop new medicines. Similarly, the challenge for agriculture is the low price responsiveness of the market on both the consumer and producer sides. As we have seen, another challenge is public policies to ensure that we always have access to a safe, abundant supply of food.

Macro Policy for the Future

Clearly, government has a role in establishing a public policy that affects the agricultural industry as a whole. These macro agricultural policies should be tailored to address the unique characteristics of the agricultural sector, meet the needs of society as a whole, and not depress farm prices in the United States or developing countries. An international program of supply management for the major crops—corn, wheat, soybeans, and rice—could be a foundational set of policies benefiting farmers worldwide. The policy would have three elements: the establishment of an international humanitarian food reserve, the institution of an acreage reduction program by the top two or three producing countries for a given crop, and a storage program to maintain prices within a predetermined range.

In order to understand why a supply management program is necessary, a look at what has happened to crop prices since 1996 is helpful. With the adoption of the 1996 Farm Bill and its radical free market approach to agricultural programs, prices for the major U.S.-produced commodities fell by as much as half of their 1995–1996 highs (Ray et al. 2001). For instance, by 1998 the price of corn was $0.45 a bushel lower than in the immediately preceding years (Schaffer 2004), soybeans were $1.09 a bushel lower, and cotton was $0.15 a pound lower (Schaffer 2003). While U.S. producers were partially shielded from the impact of these low prices by a combination of fixed payments, emergency payments, and loan deficiency payments, farmers in much of the rest of the world had to bear the brunt of lower prices without any protection.

As the oligopoly price leader in the major agricultural commodities, the U.S. nonrecourse loan rate set a floor under the market for producers of these commodities in lands around the world. Typically, small operators in an oligopolistic market price their products just under the price leader and quickly clear their markets. When the price floor was removed, prices fell, taking farmers around the world down with them. Counter to the accusations that U.S. subsidies drove U.S. production up and global prices down, it was the *decoupling* of U.S. farm payments from the nonrecourse loan program that hurt farmers worldwide. The high payments that critics (Oxfam 2002; World Bank 2003) talk about were the result of low prices, not the cause (Ray et al. 2003, 9). Again, the cause was the decoupling of U.S. payments from the nonrecourse loan program and the elimination of annual acreage reduction programs in the United States.

An international supply management program would produce benefits far beyond the circle of large country producers who receive the direct payments for participating in the program. The payments should be structured to encourage a critical mass of farmers to participate in a supply management program, while using tiered payments that favor small and midsize farmers.

Separating the humanitarian reserve from the price stabilization storage program protects it from the abuses of the past, when food aid was used as a means of stabilizing prices in the U.S. market. The price stabilization storage program would absorb excess stocks in years of good production and release these stocks into the marketplace in years of a production shortfall.

While the humanitarian reserve is clearly targeted at the need for food security, the price stabilization storage program serves that broader function as well. The experience of the years since the implementation of the 1996 Farm Bill makes it clear that counter to the expectations of the bill's proponents, the commercial sector has no financial incentive to maintain reserves to ensure the availability of a sufficient amount of grains and seeds to be able to cope with a situation in which, for instance, the United States and Brazil experienced production shortfalls of 15 percent or more in two successive years. Without both a humanitarian reserve and a price stabilization storage program, prices would skyrocket, and malnutrition, hunger, and starvation would expand well beyond the eight hundred million people currently facing these conditions.

One of the new and innovative means of addressing the need to manage crop production levels is to put a portion of U.S. farmland into the production of dedicated bioenergy crops such as switchgrass. Instead of "paying farmers not to farm"—an accusation made about acreage-reduction programs in the past—a payment could be provided to help farmers cover the cost of establishing a stand of switchgrass. As a perennial crop, switchgrass would help to reduce soil erosion and provide wildlife habitat while remaining available for conversion back to crop production, should the need arise (De La Torre Ugarte et al. 2003). The payments could be directed in ways that strengthen farming as a household livelihood strategy.

Once the necessary technological advances have been made, small farmers could be aided in organizing cooperatives to develop and operate cellulose-to-ethanol plants that could use switchgrass as one of its feedstocks. Additionally, switchgrass production could be used to temporarily take areas out of production that are facing serious disease or pest infestation. For instance, this could be important in nematode-infested fields for which a two-year corn-soybean rotation is not sufficient to reduce the nematode numbers.

The implementation of programs to support the conversion of some cropland for the production of bioenergy crops provides a means of managing the land resources of agriculture over the long term. The management of the use of land over the short to medium term could be accomplished by a continuation of the Conservation Reserve Program that serves as a medium-term tool for the management of land resources and the institution of annual set-aside programs that serve as a short-term management tool. All of these programs help support the goal of food security by maintaining the availability of land resources for the production of food.

If the combination of tools that have been identified were used judiciously and with an eye on the long-range goals of food security and farming as a household livelihood strategy, wide swings in the use of land resources could be avoided, providing overall stability for the agricultural sector. The key is the willingness of society to maintain various food storage programs that are evaluated not in terms of minimizing costs but in terms of maximizing food security and farmer household livelihood.

Summary and Conclusions

If present trends continue, the distinct possibility exists that the U.S. farming landscape will look more and more like that of many Latin

American countries, with a few latifundia surrounded by a large number of minifundia and few producers in between—a pattern at odds with the Jeffersonian ideal of the independent yeoman farmer as the backbone of U.S. democracy. This transition toward a polarized agriculture, however, need not be seen as inevitable. Public policies can be put in place to discourage the further consolidation of U.S. agriculture and encourage the growth of small to midsize operations in which farming is a household livelihood strategy. Coupled with these public policies are the private ones of numerous individual producers who are called on to constantly reassess the economic landscape and make the adaptations necessary to maintain farming as a household livelihood strategy. Adaptability, flexibility, and managerial skills are crucial to the operation as well as survival of small to midsize family farming operations.

One of the first steps in developing both public and private policy is the recognition of the fact that the potential for oversupply and low prices is characteristic of agriculture in general along with any given differentiated product in particular. Attention has to be given to the need to find and maintain the balance point between production and demand at which producers can sustain a livelihood for themselves and their families. At the national and international levels that means the establishment of supply management programs with a number of storage programs designed to meet various needs. These programs are crucial, for instance, to those midsize producers who earn a premium for growing an intellectual property version of a commodity crop. That premium will be most beneficial only if the base price of the commodity is at a level that covers its cost of production.

One key task for small to midsize farmers is the identification of a product that can be differentiated from the commodities that are produced by the large operators and integrators. These products can range from ritually slaughtered goats prepared for the Islamic market, to soybeans produced for the edamame market, to beef raised without the prophylactic use of antibiotics. In each case, producers need to be attuned to identifying the needs of and supplying differentiated products for various groups of consumers whose needs are not being met by the industrialized, commodified marketplace.

Those who seek to produce these differentiated products will need to remain on guard against allowing overproduction that would drive the price below the cost of production. Public policies will need to be put in place to provide a variety of organizational forms that producers can use to coordinate the production, processing, and marketing of their products. The land grant triad of education, research, and extension needs to direct more of its resources toward meeting the needs of the producers who seek to engage in farming as a household livelihood strategy.

References

Cochrane, Willard W. 2003. *The Curse of American Agricultural Abundance: A Sustainable Solution*. Lincoln: University of Nebraska Press.

Cronon, William. 1991. *Nature's Metropolis: Chicago and the Great West*. New York: W. W. Norton.

Dagher, Magid. 2004. Personal communication.

De La Torre Ugarte, Daniel G., Marie E. Walsh, Hosein Shapouri, and Stephen P. Slinsky. 2003. The Economic Impacts of Bioenergy Crop Production on U.S. Agriculture. ERS Agricultural Economic Report no. 816. Washington, DC: U.S. Department of Agriculture.

Freeman, Laura. 2005. Available at ⟨http://www.laurasleanbeef.com/⟩.

Glynwood Center. 2005. Available at ⟨http://www.glynwood.org/⟩.

Halcrow, Harold G., Robert G. F. Spitze, and Joyce E. Allen-Smith. 1994. *Food and Agricultural Policy: Economics and Politics*. 2nd ed. New York: McGraw-Hill.

Hendrickson, Mary, and William Heffernan. 2004. Can Consolidated Food Systems Achieve Food Security. Unpublished manuscript.

Lobao, Linda M., and Katherine Meyer. 2001. The Great Agricultural Transition: Crisis, Change, and Social Consequences of Twentieth Century U.S. Farming. *Annual Review of Sociology* 27:103–124.

Loughmiller, William. 2005. Personal communication. January 26. AgPro International, Inc.

Lyson, Thomas A., and Amy Guptill. 2004. Commodity Agriculture, Civic Agriculture, and the Future of U.S. Farming. *Rural Sociology* 69:370–385.

Maytag Dairy Farms. 2005. Available at ⟨http://www.maytagblue.com/⟩.

Mishra, Ashok K., Hisham S. El-Osta, Mitchell J. Morehart, James D. Johnson, and Jeffrey W. Hopkins. 2002. Income, Wealth, and the Economic Well-being of Farm Households. ERS Agricultural Economic Report no. 812. Washington, DC: U.S. Department of Agriculture.

Organic Valley Family of Farms. 2005. Available at ⟨http://www.organicvalley.coop/⟩.

Ostrom, Marcia, and Douglas Jackson-Smith. 2005. Defining a Purpose: Diverse Farm Constituencies and Publicly Funded Agricultural Research and Extension. *Journal of Sustainable Agriculture* 27 (3): 57–76.

Oxfam International. 2002. Cultivating Poverty: The Impact of U.S. Cotton Subsidies on Africa. Available at ⟨http://www.oxfam.org.uk/what_we_do/issues/trade/downloads/bp30_cotton.pdf⟩.

Pirog, Rich. 2004. A Geography of Taste: Iowa's Potential for Developing Place-Based and Traditional Foods. Ames: Iowa State University. Available at ⟨http://www.leopold.iastate.edu/pubs/staff/taste/taste.htm⟩.

Ray, Daryll E. 2001a. How Did the 1996 Farm Bill Come to Be: Long-Term Influences. *MidAmerica Farmer Grower*, January 12. Available at ⟨http://www.agpolicy.org/articles01.html⟩.

Ray, Daryll E. 2001b. Total Crop Demand Increases Very Little When Prices Decline, Contributing Little to Market Self-Correction. *MidAmerica Farmer Grower*, April 6. Available at ⟨http://www.agpolicy.org/articles01.html⟩.

Ray, Daryll E. 2001c. Disconnection between Agricultural Production and Consumption Needs. *MidAmerica Farmer Grower*, December 14. Available at ⟨http://www.agpolicy.org/articles01.html⟩.

Ray, Daryll E., Daniel G. De La Torre Ugarte, and Kelly Tiller. 2003. *Rethinking U.S. Agricultural Policy: Changing Course to Secure Farmer Livelihoods Worldwide*. Knoxville: Agricultural Policy Analysis Center, University of Tennessee.

Ray, Daryll E., Mozhgan Shahidi, and Harwood Schaffer. 2001. *An Analytical Database of U.S. Agriculture, 1950 to 1999: The APAC Database, 2001*. Agricultural Policy Analysis Center Staff Paper no. 01–1. Knoxville: University of Tennessee Agricultural Experiment Station.

Schaffer, Harwood D. 2003. Unpublished econometric models on soybeans and cotton.

Schaffer, Harwood D. 2004. On Predicting the Price of Corn, 1963–2002. *Journal of Agricultural and Applied Economics* 56 (2): 520.

Shannon, Fred A. 1989. *The Farmer's Last Frontier, Agriculture, 1860–1897*. Vol. 5 of *The Economic History of the United States*. Reprint, Armonk, NY: M. E. Sharpe.

Stokes, Fred. 2004. Personal communication.

Tiller, Kelly. 2000. Contracting in U.S. Agriculture: Lessons for Tobacco. PowerPoint presentation. Available at ⟨http://apacweb.ag.utk.edu/ppap/pp00/contract.pdf⟩.

World Bank. 2003. *Global Economic Prospects: Realizing the Development Promise of the Doha Agenda*. Washington, DC: World Bank. Available at ⟨http://web.worldbank.org/external/default/main?theSitePK=544296&contentMDK=20281499&menuPK=544304&pagePK=64167689& piPK=64167673⟩.

9

Agriculture of the Middle: Lessons Learned from Civic Agriculture

Thomas A. Lyson

Agriculture and food production in the United States are being restructured along two distinct lines. On the one hand, large-scale, management- and capital-intensive, technologically sophisticated, industrial-like farm operations are becoming tightly tied into a network of national and global food purveyors. These farms will produce large quantities of highly standardized bulk commodities. A few hundred large farms will account for most of gross agricultural sales, but not necessarily farm income.

On the other hand, a substantial number of smaller-scale, locally oriented, flexibly organized farms and food producers are taking root throughout the United States. These farms are part and parcel of what I call the new civic agriculture (Lyson 2000, 2004). And if the current trends continue, civic agriculture will likely expand in scope to become an enduring feature of the agricultural landscape. These farms and food processors along with their attendant markets will fill the geographic and economic spaces that have been passed over or ignored by large-scale, industrial producers.

The agriculture of the middle farms are being squeezed by the emerging bifurcation of production. On the one hand, economies of scale dictate that markets for basic commodities favor large farmers over midsize producers. On the other hand, agriculture of the middle producers are ill suited to provide the specialty products and serve the niche markets that are being exploited by civic agriculture farmers.

Civic agriculture is the embedding of local agricultural and food production in the community. Civic agriculture is not only a source of family income for the farmer and food processor; civic agricultural enterprises contribute to the health and vitality of communities in a variety of social, economic, political, and cultural ways (DeLind 2002). For example, civic agriculture increases agricultural literacy by directly linking consumers to producers. Likewise, civic agricultural enterprises have a much higher local economic multiplier than farms or food processors that are producing for the global mass-market. Dollars spent for locally produced agricultural and food products circulate several more times through the local community than the money spent for food products that are processed and packaged by multinational corporations and sold in national supermarket chains.

Civic Agriculture and Agriculture of the Middle

To more fully understand the processes and structures associated with civic agriculture, it is useful to compare civic agriculture to the conventional agriculture paradigm. The conventional approach to farming is sometimes referred to as commodity agriculture or the productivist paradigm (see Lappe and Lappe 2002), and it encompasses both the large-scale industrial farm sector as well as the traditional agriculture of the middle producer. Conventional agriculture is grounded on the belief that the primary objectives of farming should be to produce as much food/fiber as possible for the least cost. It is driven by the twin goals of productivity and efficiency (Lyson and Welsh 1993).

Conventional agriculture is anchored to a scientific paradigm that is rooted in experimental biology, and embodies an approach to farming that focuses on enhancing the "favorable" traits of crop varieties and animal species. Further, in capitalist economies, the products developed by agricultural scientists are turned into commodities that can be bought, sold, and traded on the world market. As such, the reductionist nature of experimental biology, which identifies/creates "traits," dovetails nicely with the reductionism of neoclassic economics, which provides the framework for turning these traits into "property."

The conventional model of agriculture concentrates primarily on commodities and their component parts as units of observation, analysis, experimentation, and intervention. Farmers and farms have largely been ignored by the conventional agriculture community. Indeed, farmers are often reduced to *workers* whose primary tasks are to follow production procedures outlined from above. And farms are simply *places* where production occurs devoid of any connections to the local community or social order.

The primary advocates for the conventional model of farming in the United States have been the land grant colleges and universities, various agencies within the USDA, and more recently large, multinational agribusiness firms. The land grant system was organized to introduce and disseminate to students and farmers alike the methods of scientific research in agriculture. At U.S. land grant universities, the emphasis in the classroom and research laboratory has been *production*. As different production-oriented agricultural disciplines were formed over the past 120 years such as agronomy, plant pathology, the animal sciences, plant breeding, and entomology, they broke apart "farming" bit by bit into disciplinary niches.

Across disciplines, the goals were the same. In the plant and soil sciences, attention was directed at increasing yields by enhancing soil fertility, reducing pests, and developing new genetic varieties. Animal scientists, conversely, focused on health, nutrition, and breeding. The scientific and technological advances wrought by land grant scientists were filtered through a farm management paradigm in agricultural economics that championed sets of "best management practices" as the blueprints for successful and presumably profitable operations.

The conventional approach to production agriculture treats the farm operator as a manager and an individual "problem solver" (Kay 1986). The role of the USDA's Cooperative Extension Service, which is still the primary educational outreach organization for farmers, has been to supply producers with the knowledge, skills, and information necessary to make the best decisions within the parameters of their own farms. The individual not the community has been the sole locus of attention for program development and outreach efforts. Farmers who "failed" to make a profit and subsequently went out of business, whether or not they followed the prescriptions of the Cooperative Extension Service, were deemed "bad managers."

Industrialization is the motor behind the production model of agriculture. According to Rick Welsh (1996, 3), "Industrialization has traditionally referred to the process whereby agricultural production has become less of a subsistence activity and more of a commercial activity." In the United States, agricultural industrialization has proceeded relatively unabated since the 1920s. Farms have become larger in size and fewer in number, especially after World War II, when large amounts of chemical fertilizers and synthetic pesticides became part of their standard operating practices. Land has been used more intensively and yields per acre of farmland have increased dramatically. The amount of farmland has decreased while capital investment on the farm has increased. At the same time, farms have been woven into ever-tighter marketing channels. One consequence of industrialization is that over time, it allowed a group of large producers to dominate different sectors of the agricultural economy. Consider that there were 53,641 farms that reported vegetable sales in the 1997 Census of Agriculture. But the largest 3,074 (5.8 percent of the total) of these farms—those with annual sales of $1 million or more—accounted for 75.4 percent of all vegetable sales in the country. On average, these large vegetable farms sold about $2.1 million every year.

Over the past twenty-five years, several problems associated with the large-scale industrial model of farming have become apparent. These problems center on both the environmental and social or community aspects of industrial agriculture. For example, it is an acknowledged fact that industrial agriculture had become resource intensive. Many industrial agricultural techniques exploited the land, polluted water, and used large amounts of energy. In many ways, green revolution technologies became the epitome of the industrial agriculture model (Conway 1997).

Beyond issues of environmental degradation and resource depletion, however, are the negative social and community consequences associated with industrial agriculture. Because the industrial model is anchored to the neoclassical theory of economics, it is not surprising that this model neglects entirely issues of rural community and farm viability. By fixing attention solely on production and efficiency, the industrial model does not take into account farm household and community welfare. Yet there is abundant empirical evidence that the structure of agriculture affects community well-being. In a seminal study undertaken in California, Walter Goldschmidt (1978) showed that communities in which the economic base consisted of many smaller, locally owned farms (i.e., agriculture of the middle farms), manifested higher levels of social, economic, and political welfare than those where the economic base was dominated by a few large absentee-owned farms. Paul Durrenberger and Kendall Thu (1996) found similar patterns in Iowa. Research on U.S. counties by Linda Lobao (1990) and Lyson and colleagues (2001), among others, has also confirmed Goldschmidt's findings. The ongoing "farm crisis," especially in North America, is characterized by low commodity prices, a lack of competitive markets, and the abnegation of responsibility by governments for the welfare and well-being of farmers and their communities.

Civic Agriculture and Problem Solving

Civic agriculture may have something important to offer agriculture of the middle producers: it is fundamentally about problem solving. As such, it represents a substantially different approach to farming and food production than the industrial one to growing crops as well as raising livestock and poultry. Civic agriculture rests on a biological paradigm best described as "ecological," and hence is not readily amenable to incorporating the techniques or technologies of reductionist science. Ecological approaches to agriculture seek not so much to increase output or yield but to identify and moderate production processes that are "optimal."

Smaller-scale agricultural and food ventures that are tied to the community through direct marketing or integrated into local circuits of food processing and procurement embody the civic concept. Taken together, the enterprises that make up and support civic agriculture can be seen as part of a community's *problem-solving capacity* (Young 1999). The locally based organizational, associational, and institutional component of the agriculture and food system is at the heart of civic agriculture. Local producer and marketing cooperatives, regional trade associations, and community-based farm and food organizations are part of the underlying structure that supports civic agriculture, and could also bolster the agriculture of the middle.

Civic agriculture is a locally organized system of agriculture and food production characterized by networks of producers who are bound together by place. It embodies a commitment to developing and strengthening an economically, environmentally, and socially sustainable system of agriculture and food production that relies on local resources, and serves local markets and consumers. The imperative to earn a profit is filtered through a set of cooperative and mutually supporting social relations.

Civic agriculture should not be confused with civic farmers. Farmers who vote in local elections, sit on school boards, are active members of local service clubs such as the Rotary, Lions, or Kiwanis, and otherwise participate in the civic affairs of their communities may be seen as "good citizens." Nevertheless, a farm or food business that is not woven into the social and economic fabric of the local community, produces only for the export market, relies on nonlocal hired labor, and provides few benefits for its workers is not a civic enterprise, regardless of the civic engagement of its operator.

Obviously, no agricultural or food enterprise is without some civic merit. Yet large-scale, contract poultry and hog operations, farmers who sell only to large food corporations such as Tyson, Perdue, or Hormel, would lie at the far outside end of civic-mindedness. Likewise, large-scale, absentee-owned or operated industrial-like fruit and vegetable farms that rely on large numbers of migrant workers and sell their produce for export around the world would not be deemed civic.

Civic Agriculture, Agriculture of the Middle, and Sustainable Agriculture

Sustainable agriculture is a term that became popular in the 1980s as an organized response to many of the shortcomings of the industrial model of agricultural production. The term sustainable agriculture has come to denote a more environmentally sound and socially responsible system of agricultural production than has traditionally existed in most Western societies. While there are many definitions of sustainable agriculture, one of the more widely accepted ones was introduced by the USDA and published as part of the 1990 Farm Bill. According to the USDA (FACTA 1990), sustainable agriculture is

an integrated system of plant and animal production practices having a site-specific application that will, over the long-term: 1) satisfy human food and fiber needs; 2) enhance environmental quality and the natural resource base upon which the agricultural economy depends; 3) make the most efficient use of non-renewable resources and integrate, where appropriate, natural biological cycles and controls; 4) sustain the economic viability of farm operations; and 5) enhance the quality of life for farmers and society as a whole.

It is important to note that this definition contains economic, environmental, and social or community dimensions. Sustainable agriculture encompasses a set of production practices that are economically profitable for farmers, preserve and enhance environmental quality, contribute to the well-being of farm households, and nurture local community development. Sustainable agriculture denotes a holistic, systems-oriented approach to farming that focuses on the interrelationships of social, economic, and environmental processes.

In an important study of the differences between large-scale industrial agriculture and sustainable agriculture, Curtis Beus and Riley Dunlap (1990) identified key elements that distinguish the two agricultural paradigms. Beus and Dunlap's key results are summarized in table 9.1. They saw the domination of nature versus harmony with nature as one of the crucial points of difference between the two approaches. Likewise, the reductionist nature of industrial agriculture was captured by the emphasis on commodity specialization, while the problem-solving attribute of sustainable agriculture was aligned with diversity.

The underlying social science paradigms were portrayed by Beus and Dunlap as competition versus community. Industrial agriculture rests on a business orientation, with a primary stress on speed, quantity, and profit. The community orientation of sustainable agriculture, on the other hand, rests on cooperation, with an emphasis on permanence, quality, and beauty.

It is not too difficult to see the connections between sustainable and civic agriculture, and by extension to agriculture of the middle producers. Indeed, sustainable agriculture can be viewed as a logical antecedent to civic agriculture. The term civic agriculture captures the problem-solving foundations of sustainable agriculture. But civic agriculture goes further by referencing the emergence as well as growth of community-based agriculture and food production activities that create jobs, encourage entrepreneurship, and strengthen community identity. Civic agriculture brings together production and consumption activities within communities, and offers real alternatives to the commodities produced, processed, and marketed by large agribusiness firms (Lyson and Green 1999; Green and Hilchey 2002).

Table 9.1
Selected Elements of Industrial Agriculture and Sustainable Agriculture

Industrial agriculture	Sustainable agriculture
Domination of nature	*Harmony with nature*
Humans are separate from and superior to nature	Humans are part of and subject to nature
Nature consists primarily of resources to be used	Nature is valued primarily for its own sake
Life cycle incomplete; decay (recycling of wastes) neglected	Life cycle complete; growth and decay balanced
Human-made systems imposed on nature	Natural ecosystems are imitated
Production maintained by agricultural chemicals	Production maintained by development of healthy soil
Highly processed, nutrient-fortified food	Minimally processed, naturally nutritious food
Specialization	*Diversity*
Narrow genetic base	Broad genetic base
More plants grown in monocultures	More plants grown in polycultures
Single cropping in succession	Multiple crops in complementary rotations
Separation of crops and livestock	Integration of crops and livestock
Standardized production systems	Locally adapted production systems
Highly specialized, reductionist science and technology	Interdisciplinary, systems-oriented science and technology
Competition	*Community*
Lack of cooperation, self-interest	Increased cooperation
Farm traditions and rural culture outdated	Preservation of farm traditions and rural culture
Small rural communities not necessary to agriculture	Small rural communities essential to agriculture
Farmwork a drudgery; labor input to be minimized	Farmwork rewarding; labor an essential to be made meaningful
Farming is a business only	Farming as a way of life as well as a business
Primary emphasis on speed, quantity, and profit	Primary emphasis on permanence, quality, and beauty

Source: Adapted from Beus and Dunlap 1990.

Profiling Civic Agriculture

Industrial agriculture produces most of the food and fiber in the United States. Nevertheless, smaller-scale, locally oriented producers and processors have become key actors in revitalizing rural areas. These producers represent the vanguard of a social trend toward civic agriculture and reviving the agriculture of the middle.

To be sure, there is an emerging debate about the likelihood that civically organized local food systems and the agriculture of the middle can continue to expand as well as flourish in a globalizing environment. Over the past ten years, however, an accumulating body of research has begun to assess the benefits of small and midsize enterprises at the level of civic and community welfare (Piore and Sabel 1984; Lyson et al. 2001). Communities that nurture local systems of agricultural production and food distribution as one part of a broader plan of economic development may gain greater control over their economic destinies, and contribute to rising levels of civic welfare and socioeconomic well-being.

Civic agriculturalists and farmers of the middle along with their enterprises are a varied lot, and no one set of characteristics perfectly defines these producers. Yet a profile of civic operations compared to industrial agricultural producers can be constructed. First, civic agriculture is oriented toward local market outlets that serve local consumers rather than national or international mass markets. Civic agriculture is seen as an integral part of rural communities, not merely as the production of commodities. The direct contact between civic farmers and consumers nurtures bonds of community. In civic agriculture, producers forge direct market links to consumers rather than indirect links through middlepeople (wholesalers, brokers, processors, etc.).

Second, farmers engaged in civic agriculture enterprises are concerned more with high-quality and value-added products, and less with quantity (yield) and least-cost production practices. Civic farmers cater to local tastes, and meet the demand for varieties and products that are often unique to a particular region or locality.

Third, civic agriculture at the farm level is frequently more labor intensive and land intensive, and less capital intensive. Civic farm enterprises tend to be considerably smaller in scale and scope than industrial producers. Civic farming is a craft enterprise as opposed to an industrial one. As such, it harks back to the way in which farming was organized in the early part of the twentieth century. Civic agriculture takes up social, economic, and geographic spaces not filled (or passed over) by industrial agriculture.

Finally, civic agriculture often relies more on indigenous, site-specific knowledge and less on a uniform set of best management practices. The industrial model of farming is characterized by the homogenization and standardization of production techniques. The embedding of civic agriculture in the community along with a concern with environmental conditions fosters a problem-solving perspective that is site specific and not amenable to a one-size-fits-all mentality.

Civic agriculture enterprises are visible in many forms on the local landscape (Green and Hilchey 2002), and can be seen as part of local value chains that create jobs and generate income that remains in the community. *Farmers' markets* provide immediate, low-cost, direct contact between local farmers and consumers, and are an effective economic development strategy for communities seeking to establish stronger local food systems. *Community* and *school gardens* supply fresh produce to underserved populations, teach food production skills to people of all ages, and contribute to agricultural literacy. *Small-scale organic farmers* across the country have pioneered the development of local marketing systems and formed "production networks" that are akin to manufacturing industrial districts. *CSA* operations forge direct links between nonfarm households and their CSA farms. New *grower-controlled marketing cooperatives* are forming, especially in periurban areas, to more effectively tap emerging regional markets for locally produced food and agricultural products. *Agricultural districts* organized around particular commodities (such as wine, cheese, or fruit) have served to stabilize farms and farmland in many areas of the country. *Community kitchens* provide the infrastructure and technical expertise necessary to launch new food-based enterprises. *Specialty producers* and *on-farm processors* of products for which there are not well-developed mass markets (deer, goat and sheep cheese, free-range chickens, organic dairy products, etc.) as well as *small-scale, off-farm, local processors* add value in local communities and offer markets for farmers who cannot or choose not to produce bulk commodities for the mass market. What these civic agriculture efforts share is that they have the potential to nurture local economic development, maintain diversity and quality in products, and provide forums where civic farmers and food citizens can come together to solidify bonds of community. For agriculture of the middle producers, new mid-level value chains offer the same potential to solidify and build on the existing structure of mid-level producers.

Civic Agriculture and Food Citizenship

A consolidated, corporate-controlled food and agricultural system is able to supply vast quantities of standardized fare. The foundation of this system rests on a set of large farms articulating with a small number of global food processors, which in turn link with another small number of large and increasingly global food retailers. For the system to run *efficiently*, it must standardize and rationalize both the production and transaction costs all along the food chain. The smaller the number of players in the system, the easier it is to standardize and rationalize.

Since 1990, the retail sector of the food industry has seen the emergence of five major players: Kroger, Wal-Mart, Albertsons, Safeway, and Ahold USA. These five firms now account for over 40 percent of food retail sales. As Kaufman (2000, 21) notes, though, "The effect of consolidation on consumers is related primarily to increases in local market concentration—the combined sales of the largest firms expressed as a share of the total local market sales." The average market concentration of the top four retailers in individual metropolitan areas stands at about 75 percent. And there are several metropolitan areas in which the top four firms account for 90 percent or more of sales.

Almost all of the large food retailers cite lower costs and efficiencies as the primary benefits of consolidation. As size increases, procurement, marketing, and distribution costs all presumably decrease. Further, to lower their operating costs, the large food retailers are centralizing management and control at their headquarters. Finally, by forming supply chains with global processors, the large retailers are able to streamline the procurement and distribution of products.

A food system dominated by a small handful of large corporations offers consumers little real choice. Innovation in these firms is linked to devising better marketing strategies for a narrow range of "basic" products (i.e., soft drinks, breakfast cereals, snack and convenience foods, and the like). The fact that over twelve thousand new products are introduced each year (most of which fail) suggests that the food industry is not responding to consumer demand but rather blindly offering consumers sets of repackaged, reformulated, and reengineered products in hopes that a few will turn out to boost corporate profits (Lyson and Raymer 2000).

What does all of this mean for the consumer? The consolidation of the food industry clearly means less competition and lower costs for the producer. But recent research suggests that the consumer may not benefit. A study by Azzeddine Azzam and colleagues (2002) shows that a concentration in the food-processing industry has resulted in higher consumer prices in most sectors due in part to the ability of oligopolies to set prices.

Creating a civic agriculture and preserving the agriculture of the middle has the potential to transform individuals from passive consumers into active food citizens. A food citizen is someone who not only has a stake but also a voice in how and where their food is produced and processed. Civic agriculture and agriculture of the middle, as two aspects of the civic community, become a powerful template around which to build non- or extramarket relationships between persons, social groups, and institutions that have been distanced from each other (Kloppenburg et al. 1996).

Civic engagement with the food system is taking place throughout the country as citizens and organizations grapple with providing food for the hungry, establishing community-based food businesses, developing community and school gardens, organizing food policy councils, and linking "consumers" to "producers" through farmers' markets, u-pick operations, farm stands, and the like. While diverse, these efforts have one thing in common: they are all local problem-solving activities organized around agriculture and food.

Both the global and local food systems are works in progress. For many in the United States, the global system of agriculture and food production may seem totalizing. Yet there are alternatives forming in multiple places and at different times. Food citizenship rests on communal forms of problem solving that lead to a safe, nutritious, healthy diet, and a socially as well as environmentally sustainable system of agriculture and food production.

Note

Support for this research was provided in part by the Cornell University Agricultural Experiment Station in conjunction with USDA/CSREES (Cooperative State Research, Education and Extension Service) regional research project NE–1012. This is a revised version of a paper that was published in *Culture and Agriculture* titled "Civic Agriculture and Community Problem Solving," and it is reprinted with the permission of the publisher.

References

Azzam, Azzeddine M., Elena Lopez, and Rigoberto A. Lopez. 2002. Imperfect Competition and Total Factor Productivity Growth in U.S. Food Processing. Food Marketing Policy Center, Research Report no. 68. Storrs: University of Connecticut, Department of Agricultural and Resource Economics.

Beus, Curtis E., and Riley E. Dunlap. 1990. Conventional versus Alternative Agriculture: The Paradigmatic Roots of the Debate. *Rural Sociology* 55:590–616.

Conway, Gordon. 1997. *The Doubly Green Revolution.* London: Penguin.

DeLind, Laura B. 2002. Place, Work, and Civic Agriculture: Fields for Cultivation. *Agriculture and Human Values* 19:217–224.

Durrenberger, Paul E., and Kendall Thu. 1996. The Expansion of Large-Scale Hog Farming in Iowa: The Applicability of Goldschmidt's Findings Fifty Years Later. *Human Organization* 55:409–415.

Food, Agriculture, Conservation, and Trade Act of 1990 (FACTA). 1990. Public Law 101–624, title XVI, subtitle A, section 1603. Washington, DC: Government Printing Office. ·

Goldschmidt, Walter. 1978. *As You Sow.* Montclair, NJ: Allanheld, Osmun.

Green, Joanna, and Duncan Hilchey. 2002. *Growing Home: A Guide to Reconnecting Agriculture, Food, and Communities.* Ithaca, NY: Community Food and Agriculture Program, Department of Development Sociology, Cornell University.

Kaufman, Philip R. August 2000. Consolidation in Food Retailing: Prospects for Consumers and Grocery Suppliers. Agricultural Outlook, Economic Research Service, USDA.

Kay, Ronald D. 1986. *Farm Management: Planning, Control, and Implementation.* New York: McGraw-Hill.

Kloppenburg, Jack, J. Hendrickson, and G. W. Stevenson. 1996. Coming into the Foodshed. *Agriculture and Human Values* 13:33–42.

Lappe, Frances Moore, and Anna Lappe. 2002. *Hope's Edge.* New York: Putnam. ·

Lobao, Linda M. 1990. *Locality and Inequality.* Albany: State University of New York Press.

Lyson, Thomas A. 2000. Moving Toward Civic Agriculture. *Choices* (third quarter): 42–45.

Lyson, Thomas A. 2004. *Civic Agriculture*. Medford, MA: Tufts University Press.

Lyson, Thomas A., and Judy Green. 1999. The Agricultural Marketscape: A Framework for Sustaining Agriculture and Communities in the Northeast. *Journal of Sustainable Agriculture* 15 (2–3): 133–150.

Lyson, Thomas A., and Annalisa L. Raymer. 2000. Stalking the Wily Multinational: Power and Control in the U.S. Food System. *Agriculture and Human Values* 17:199–208.

Lyson, Thomas A., Robert Torres, and Rick Welsh. 2001. Scale of Agricultural Production, Civic Engagement, and Community Welfare. *Social Forces* 80:311–327.

Lyson, Thomas A., and Rick Welsh. 1993. Crop Diversity, the Production Function, and the Debate between Conventional and Sustainable Agriculture. *Rural Sociology* 58:424–439.

Piore, Michael J., and Charles F. Sabel. 1984. *The Second Industrial Divide*. New York: Basic Books.

Welsh, Rick. 1996. The Industrial Reorganization of U.S. Agriculture. Policy Studies Report no. 6. Greenbelt, MD: Henry A. Wallace Institute for Alternative Agriculture.

Young, Frank W. 1999. *Small Towns in Multilevel Society*. New York: University Press of America.

10

The Effect of Laws That Foster Agricultural Bargaining: Apple Growers in Michigan and New York State

Shelly Grow, Amy Guptill, Elizabeth Higgins, Thomas A. Lyson, and Rick Welsh

Cooperative bargaining is thought to be one way farmers can improve their economic power when faced with concentration and vertical integration among buyers of agricultural products (Brandow 1970; Bunje 1980; Clodius 1957; Frederick 1990; Garoyan and Thor 1978; Helmberger and Hoos 1965; Hoos 1962; Iskow and Sexton 1992). Studies have long assumed that as the processing and other purchasing sectors of raw farm products consolidated and vertically integrated, farmers would form or join bargaining associations. Bargaining associations provide farmers with a tool for setting minimum prices for their products, negotiating with companies having investments in other countries, bargaining for beneficial trade agreements, gaining more detailed market and price information, and otherwise increasing their economic power (Clodius 1957; Lang 1994; Levins 2001).

One of the strongest benefits of agricultural bargaining is negotiating favorable contract terms (Brandow 1970; Bunje 1980; Helmberger and Hoos 1965; Hoos 1962). According to Ralph Bunje (1980, 105), "The fading open assembly and free markets are being replaced by a system of contract farming. Contract farming lends itself to group association by producers for negotiating contract terms." In the future, he predicted, "contractual agriculture will be bargained agriculture." Echoing Bunje's words, Randall Torgerson (1998, 1), deputy administrator of the Rural Business-Cooperative Service of the USDA, stated in 1997 that "it is likely that the need for negotiated pricing will increase in dealing with production and marketing of identity-preserved products and increased use of contracting for production and services in both the crop and livestock sectors."

Table 10.1
Contracting in U.S. Agriculture, 1978 and 1997

Year	Number of farms with contracts	Percentage of farms with contracts	Percentage of total value of commodities produced under contract
1978	43,665	1.9	10
1997	227,481	11.1	31.2

Source: Banker and Perry 1999; USDA-ERS 1996.

Nearly thirty years after Bunje's observation, contracting in U.S. agriculture has increased (table 10.1), but the development and strengthening of cooperative bargaining associations has not kept pace with changes in the nation's agricultural sector. In 1978, Mahlon Lang identified sixty-seven active associations in 13 states. Fourteen years later, Julie Iskow and Richard Sexton (1992) identified thirty-six associations in nine states. There are currently about nineteen agricultural bargaining associations active in nine states (USDA-RBS 2001).

While a variety of factors contribute to the ability of bargaining associations to form and bargain effectively, many authors contend that the most critical factor is strong laws that support the right of farmers to collectively bargain, with meaningful sanctions for violations (Iskow and Sexton 1992; Marcus and Frederick 1994; Torgerson 1998). These laws are necessary in order to prohibit major buyers from engaging in antiorganizing tactics when faced with farmers organizing. These tactics have historically included failing to renew the contracts of farmers who joined bargaining associations, requiring farmers to resign from an association as a part of their contract, offering "sweetheart deals" to farmers who withdrew from associations, and threatening farmers who attended organizing meetings (Bunje 1980).

The Agricultural Fair Practices Act of 1967 (AFPA or "the act") was passed by Congress in direct response to these antiorganizing tactics. It was a landmark piece of legislation in that it explicitly supported the concept of collective action by farmers. The AFPA states that there is a need for farmers to be free to join together voluntarily in cooperative organizations, and makes this the policy of the U.S. government (Bunje 1980). Under the AFPA, it is unlawful for any handler knowingly to do the following:

- Coerce a producer to join or refrain from joining an association, or refuse to deal with a producer because that producer has joined one
- Discriminate against a producer with respect to price quantity because of their membership in an association
- Coerce a producer to breach or terminate their association membership
- Offer an inducement to a producer to cease being an association member or refuse to join
- Make false reports about an association
- Conspire with another to commit any of the above (Frederick 1990)

It is argued, however, that flaws in the AFPA have actually hindered the formation of bargaining associations (Frederick 1993). The following disclaimer clause of the act is particularly problematic: "Nothing in this chapter shall prevent handlers and producers from selecting their customers and suppliers for any reason other than a producers' membership in or contract with an association or producers, nor require a handler to deal with an association of producers" (7 U.S.C. §2304).

This language allows buyers to refuse to have meaningful negotiations with bargaining associations—probably the most serious problem faced by bargaining cooperatives. In addition, processors have used a federal preemption clause in the AFPA to challenge state laws that require good faith negotiations and binding arbitration (Frederick 1993; Marcus and Frederick 1994).

Other provisions further weaken the ability of the AFPA to support the formation of bargaining associations. The federal law is notable for its lack of enforcement authority for violations. Injured parties are given the right to sue in U.S. district courts or file a complaint with the secretary of agriculture, who could file suit through the attorney general. Litigation is not an attractive option for most bargaining associations, however; in addition to the expense, producers prefer to have cordial relations with their buyers because there are few processors from which to choose (Marcus and Frederick 1994). Also, processors are generally in a far better position than the cooperative to survive a battle of attrition. As a result, producers have sought to bring their complaints to the USDA rather than file suit themselves. Yet enforcement of the act through the USDA has been unsatisfactory, largely due to limited personnel and a lack of funding at the agency to pursue complaints, unwillingness on the part of the attorney general to take action, and difficulty in finding producers who are willing to testify against a processor, for fear of reprisals (ibid.). In addition, the legislated penalties for violating the AFPA are so low that buyers have little incentive to comply with the law. As a result, from the time of enactment in 1967 until 1989, only approximately twenty-six complaints had been filed under the AFPA, and only about six resulted in favorable outcomes for the growers or their associations (Bunje 1980; Frederick 1990).

In the thirty years since the passage of the act, there have been numerous attempts to change the federal legislation to substantially improve the climate for cooperative bargaining associations (Frederick 1990). Some recommended changes (Bunje 1980; Frederick 1990, 1993) include the following:

- Repeal of the disclaimer clause
- Provisions for fee deductions
- Requirements for negotiators to bargain in good faith
- Provisions for mediation or arbitration
- Provisions for qualifying or otherwise accrediting a bargaining association
- Provisions for defining a bargaining unit, and providing for the designation or selection of an exclusive agent for the bargaining unit
- Adoption of "agency shop" language to eliminate "free-rider" problems
- Protective rules, and a means for promulgating and administering them
- Authority for the USDA to assess civil penalties for violations of the act
- Limit on federal preemption
- Inclusion of language to cover contract production in the poultry and livestock industry (Morrison 1995)

To date, attempts to implement these recommendations have been unsuccessful. Nine states have adopted state laws that are stronger than the federal law, and it appears to be no accident that these states are mainly those with agricultural bargaining associations. Of the nineteen active bargaining associations identified, sixteen were in states with laws regarding agricultural bargaining associations (USDA-RBS 2001). Most of these state laws require "good faith" bargaining between the handlers and associations, and some include dispute resolution mechanisms, such as conciliation, mediation, or arbitration (Frederick 1993).

Michigan's law, in particular, is often considered a prototype of what improved federal legislation should look like. Unlike the federal law, it provides for a specified period of time during which negotiations must take place, and it deals with impasse problems by requiring mediation or arbitration. Arbitrators under Michigan law must choose one of the last offers presented by the parties. The result is that both sides tend to make their final offers sufficiently reasonable to be persuasive to the chairperson (Bunje 1980).

It has been argued that the formation of cooperative bargaining associations with the power to engage in contract negotiations with processors should result in more equitable economic balance in agricultural markets, and would result in a more stable and prosperous farm sector (Frederick 1993). Anecdotally, this would seem to be the case. A 1999 *Successful Farming* magazine article credited the Michigan Agricultural Cooperative Marketing Association (MACMA) with keeping the prices of processing apples and asparagus in Michigan above national averages (Looker 1999). Yet there have been few studies that have attempted to quantify the potential difference in producer welfare made possible by having the power to bargain (Garoyan and Thor 1978; Helmberger and Hoos 1965; Hoos 1962). Furthermore, most studies on agricultural bargaining are outdated; the most recent was undertaken in 1992, and prior studies were done in the 1970s.

The conditions theorized to support the formation of successful cooperative bargaining associations currently exist—that is, the agricultural sector is seeing declining cash markets for commodities and increasing concentration in the processing sector. In addition, policymakers continue to introduce legislation that would strengthen and protect producer-organizing activity. Research is therefore needed to determine the economic and social impacts of agricultural bargaining associations in today's market. Determining the economic and social impacts of current federal legislation on the farm sector—by comparing them to the farm sector impacts of enhanced state legislation on cooperative bargaining—provides a starting point for future debates around changes in federal policy.

The primary ways to measure the impacts of bargaining associations are to look at their ability to command higher prices for their members than would be expected in the absence of an association; and examine their ability to command better fringe benefits for their members, such as more uniform or favorable contract terms. Because analyses of the economic and social benefits of cooperative agricultural bargaining associations are extremely outdated, this study offers insight into the economic and social benefits provided by bargaining associations as well as a strong legal framework of support in today's farm sector.

Methodology

In order to provide a starting point for understanding the potential impacts of strengthened cooperative bargaining laws at the federal level, this study examines the experiences of two states—one with a strengthened state law that incorporates many of the demands included in recently introduced federal legislation, and one without any law beyond the AFPA.[1] To test the hypothesis that the presence of bargaining power increases producer welfare, we analyzed price data using a method developed by Peter Helmberger and Sidney Hoos (1965) to look at the price effects of bargaining associations. To examine the fringe benefits that bargaining power gives to contracts, negotiations, and producer attitude, we conducted a survey of apple farmers in Michigan and New York State. These states and this commodity were selected for several reasons:

• Michigan has a strong law in support of bargaining associations that requires binding arbitration and good faith bargaining. New York has no such law.

• Michigan has a functioning bargaining cooperative for apples while New York has none (a previously formed cooperative in New York is now defunct).

• Apples are an important crop in both states. Michigan ranks number three in the United States in apple production while New York ranks number two. Both states process a significant portion of their apple harvest and use similar processing methods (i.e., canning, juice, and freezing).

• The apple commodity system fits the criteria outlined by Iskow and Sexton (1992) for potentially successful associations: limited ability for short-term entry into the industry (perennial fruit crop) and a concentrated processing sector. In addition, apple farmers in Michigan and New York share processors. This reduces the chance that New York farmers may fear extreme reprisals from processors should they form a bargaining association, as these processors have demonstrated their willingness to work with a bargaining association in Michigan.

Price Comparison Methods

Helmberger and Hoos (1965), who conducted the most comprehensive study to date on the effects of agricultural bargaining cooperatives, developed several different statistical tests to measure the differences in price due to an agricultural bargaining association. One approach they thought might be useful for determining the degree to which bargaining associations influenced prices was referred to as an "interperiod-intermarket analysis." This approach involves comparisons over two periods of time between the prices received by growers in one market and the prices received by growers in another market, where a bargaining association exists in the second period for only one of the markets. We duplicated Helmberger and Hoos's test as closely as possible, although limited by the lack of consistent pre-1969 data, by comparing Michigan and New York price data. Helmberger and Hoos recommended the use of an intermarket comparison between two separate markets for the same commodity, making Michigan and New York appropriate states for this test.

We collected price data on Michigan and New York processed apples, delineated by processing type, from 1969 to 2001 (USDA 1977, 1980, 1985, 1990, 1995, 1998, 1999, 2000, 2001, 2002). We were unable to use pre-1969 data because the manner in which they were collected was not consistent with later years, and did not distinguish between the overall price received for apples and the prices received for fresh versus processed apples. We converted all the prices into January 2002 dollars by making the national Consumer Price Index utilize January 2002 as its reference point and then applying the index to all price data (U.S. Department of Labor 2002). The resulting information helped to determine the effect of a strengthened state bargaining law on the prices received by growers.

Although increasing farmer prices is the primary stated goal of most bargaining associations surveyed (Iskow and Sexton 1992; Lang 1978), using price as a measure of success has some theoretical problems. Bargaining associations cannot control the supply of a given commodity, and therefore, according to economic theory, are not expected to greatly increase prices for their members over the long run (Helmberger and Hoos 1965; Hoos 1962). So while tests conducted in the 1960s and 1970s showed some, although not drastic, price benefits (Garoyan and Thor

1978; Helmberger and Hoos 1965), and our study also provides information on price impacts, fringe benefits are also an important service of bargaining associations.

Fringe benefits, particularly the degree of uniformity and the quality of contract terms, are in many ways the most critical measure of success of bargaining associations. Most authors agree that it is in the area of fringe benefits that bargaining cooperatives offer the most advantage to their members over the long run, especially in concentrated markets. Yet because fringe benefits are not easily quantifiable, they have not been extensively studied (Brandow 1970; Helmberger and Hoos 1965; Hoos 1962).

To gauge potential fringe benefits received by farmers, we surveyed apple farmers in Michigan and New York State to distinguish differences in the welfare and outlook of apple farmers that could be attributed to the presence or absence of the state law protecting bargaining associations (see appendix 10.A). We surveyed three groups of farmers: farmers in Michigan in the bargaining association; farmers in Michigan not in the bargaining association; and farmers in New York State.

Since many factors, such as farm size, farmer age, and a farmer's relationship with the surrounding community, play into a farmer's perceptions about their welfare and likely future in farming, we asked a set of control questions in the survey to see if there was any significant difference among the three farmer groups on what was assumed to be a set of common factors. These questions included the size of the farm, the varieties produced, the percentage processed (average), the age of farmer, the number of years in apple farming, and education level.

Because an earlier study (Babb et al. 1969) suggested that attitudes toward cooperative bargaining as well as grower and processor perceptions influenced the outcome of negotiations, some of our survey questions were meant to gauge attitudes. We asked growers a set of questions about their perception of the future of apple farming in their state, whether or not they have been a member of a bargaining association, their perception of the power that bargaining associations have in their state and to what extent the law influences that power, their relationship with their processors, and their general contract terms. We also asked apple growers about the number of processors to which they believed they had access.

Survey Methods

A total of 1,169 short mail surveys were sent to all known apple growers in Michigan (482 surveys) and New York (687 surveys). The New York State Horticultural Society and MACMA provided the names and addresses of apple growers. The first survey mailing was followed a week later by a reminder card. Two weeks after the initial mailing, a second survey was sent to nonrespondents. Of the 1,169 surveys mailed, 47 were undeliverable and returned by the U.S. Postal Service, and 122 were returned uncompleted by growers who had gone out of business or were not selling apples to processors. A total of 453 valid surveys were returned, representing 46 percent of the remaining 1,000 surveys. New York growers returned 238 surveys, or 42 percent of the remaining 589, and Michigan growers returned 215, or 52 percent of the remaining 411. About two-thirds of the Michigan respondents (143) are members of MACMA.

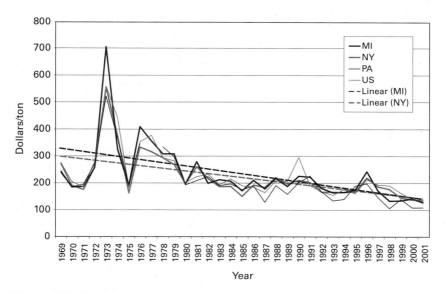

Figure 10.1
Prices for All Processed Apples (Those Canned, Frozen, and/or Used for Juices) in Michigan, New York, Pennsylvania, and the United States
Note: Prices are adjusted to 2002 prices. Trend lines indicate that prices were historically higher for Michigan apples, but that prices have converged over time. Breaks in the lines indicate that no data were available for that time period.

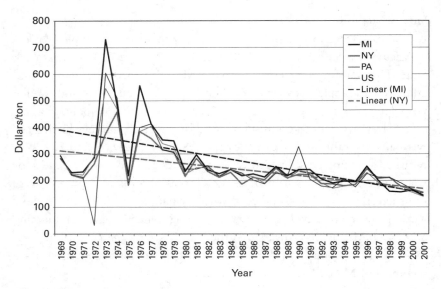

Figure 10.2
Prices for Canned Apples in Michigan, New York, Pennsylvania, and the United States
Note: Prices are adjusted to 2002 prices. Trend lines indicate that prices were historically higher for Michigan apples, but that prices have largely converged over time.

Price Comparison Results

In general, Michigan prices for processed apples ran above prices in other states, but over the years the prices have tended to converge (figures 10.1–10.3). This makes sense as the market concentrates and imported apples, especially from China, become increasingly important. Any price effect from MACMA seems to be eroding. Perhaps the more interesting story is the falling real prices for producers across the board.

Survey Results

Appendix 10.A is a table of the survey results comparing MACMA growers with Michigan non-MACMA growers and New York growers. This comparison was made to discern whether differences among growers are due to geographic distance or membership in the collective bargaining unit. Although both farm structure and attitudinal questions are included in this appendix, for the purposes of this summary we concentrate on the differences in attitude and perceptions of the three grower

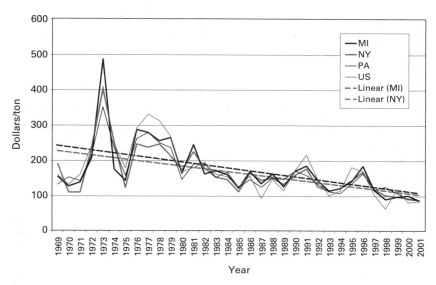

Figure 10.3
Prices for Juice Apples in Michigan, New York, Pennsylvania, and the United States
Note: Prices are adjusted to 2002 prices. Trend lines indicate that prices were historically higher for Michigan apples, but prices have converged over time.

groups. If the Michigan non-MACMA growers and the New York growers tend to answer attitudinal and perception questions more similarly to each other than to MACMA growers, we can assume that membership in the collective bargaining unit is having an effect.

Considering some of the descriptive statistics, in general MACMA growers were more likely to believe they had input into the prices received for their processed apple crop. For this group, 26 percent either agreed or strongly agreed with the statement. Only 7 percent of Michigan non-MACMA and 8 percent of New York growers agreed or strongly agreed with the statement. In addition, 18 percent of MACMA growers, 10 percent of Michigan non-MACMA growers, and 8 percent of New York growers agreed or strongly agreed that they had input into the terms of trade for their processed apple crop. Clearly, MACMA growers perceived they had some input into crucial aspects of the processed apple contracts.

MACMA growers were also more likely to believe they had input into public policy that might affect them. Almost 30 percent of MACMA growers agreed or strongly agreed with the statement that they had input into such state government policies. Less than 20 percent of the Michigan non-MACMA growers and about 25 percent of New York growers agreed with the statement. Regarding input into federal policies, 22 percent of MACMA growers, 17 percent of Michigan non-MACMA growers, and 14 percent of New York growers at least agreed that they had input. MACMA growers were also more likely than the other two grower groups to at least agree that they were generally satisfied with their marketing arrangements.

Other potentially key differences include:

• MACMA growers were more likely to desire a new federal law that requires processing firms to bargain with accredited grower bargaining cooperatives or associations. Nevertheless, 49 percent of Michigan non-MACMA and 56 percent of New York growers also at least agreed with the statement.

• MACMA growers were much more likely to agree or strongly agree with the statement that they could find marketing assistance if they needed it.

• MACMA growers were more likely to agree or strongly agree that grower bargaining units raise prices for *all* growers. Yet more than 50 percent of the growers in the other two groups also at least agreed with this statement.

To further investigate the potential impact of membership in MACMA on attitudes and perceptions, we constructed a "satisfaction" index from six survey items. These six items were the following:

• I have input into prices received for my processed apple crop
• I have input into terms of trade for my processed apple crop
• I have input into state government policies and programs that might affect my operation
• I have input into federal government policies and programs that might affect my operation
• I am generally satisfied with the marketing arrangements for my processed apple crop
• If I need it, I can get assistance in finding a "home" or market outlet for my processed apple crop

We performed a reliability analysis to determine if these six items measured a single construct that we label satisfaction. The analysis resulted in a Cronbach's alpha score of 0.77. Scores above 0.70 are generally considered to indicate that the scale is measuring a single construct.

The satisfaction scores were included in an Ordinary Least Squares Regression (OLS) analysis. Regression is a statistical technique that enables researchers to consider relationships between two variables after accounting for the effects of other factors. For this analysis, we investigated whether MACMA members were more likely to have higher satisfaction scores even after taking into account the following characteristics:

• The state in which the grower operates ($0 =$ NY; $1 =$ MI)
• Gross processed apple sales volume in bushels ($1 = 10,000$ or less; $2 = 10,001$ to $20,000$; $3 = 20,001$ to $40,000$; $4 = 40,001$ to $60,000$; $5 = 60,001$ to $80,000$; $6 = 80,001$ to $100,000$; $7 = 100,001$ to $200,000$; $8 = 200,001$ to $300,000$; $9 = 300,001$ to $500,000$; $10 = 500,001$ to $750,000$; $11 = 750,001$ to 1 million; $12 =$ more than 1 million)
• Number of years growing apples ($1 = 15$ years or less; $2 = 16$–30 years; $3 =$ more than 30 years: categories were used since some operators wrote the number of years the family had grown apples)
• The number of processing firms serving as potential buyers for the apple crop, from which the grower can choose
• Percentage of the farm's apple production sold on the processed market
• How often the grower sells to a processing firm without an agreed-on price ($1 =$ often; $2 =$ sometimes; $3 =$ seldom or never)

Table 10.2
Mean and Standard Deviation of Selected Variables

Variable	Mean	Standard deviation
Member of collective bargaining unit ($0 =$ no; $1 =$ yes)	0.31	0.46
State ($0 =$ NY; $1 =$ MI)	0.49	0.50
Gross processed apple sales volume (bushels)	4.46	2.50
Number of years growing apples	2.27	0.69
Number of processing firms	3.57	2.31
Percentage sold on processed market	58.32	29.80
Sells to processing firm without agreed-on price ($1 =$ often; $2 =$ sometimes; $3 =$ seldom or never)	2.23	0.81
Percentage of packinghouse culls	18.32	15.73
Owns enough stock to meet needs ($0 =$ no; $1 =$ yes)	0.11	0.31

• The percentage of the grower's fresh market apples that were packing-house culls.
• Whether the grower owns enough stock in a processing firm to meet all their processed apple needs (0 = no; 1 = yes)

By "controlling" for these eight farm and farmer characteristics, we can be more confident that any relationship we measure between membership in MACMA and satisfaction is genuine.

Means of the characteristics or variables are presented in table 10.2. The results from the OLS regression are presented in table 10.3. The satisfaction scale is coded such that lower scores indicate higher levels of satisfaction. For the purposes of this report, however, we reversed the signs of the coefficients such that a positive sign indicates a positive relationship with grower satisfaction.

The OLS results indicate that membership in MACMA has a significant positive effect on satisfaction, holding the other independent variables constant. Other factors that are significantly (p value ≤ 0.05) related to satisfaction include higher numbers of processing firms available as buyers, lower percentages of the apple crop sold as processed apples, fewer times selling to processing firms without an agreed-on price, lower percentage of the fresh apple crop from packinghouse culls, and owning enough stock in an apple-processing firm to meet a grower's marketing needs.

Table 10.3
OLS Results with Satisfaction-Dependent Variable

Variable	Standardized coefficient	Significance
Member of collective bargaining unit	.225	.000
State	.098	.118
Gross processed apple sales volume	.077	.092
Number of years growing apples	−.077	.064
Number of processing firms	.111	.012
Percentage sold on processed market	−.102	.024
Sells to processing firm without agreed-on price	.245	.000
Percentage of packinghouse culls	−.149	.001
Owns enough stock to meet needs	.096	.029

Note: Adj. R-square = 0.21.

The results indicate that membership in MACMA brings with it more satisfaction as measured by the scale variables. Specifically, membership in MACMA is associated with feeling generally satisfied with marketing arrangements, being able to receive help in finding apple markets, and feeling that one has input into contract terms and price as well as input into state and federal policies. These results are not surprising since MACMA actively lobbies on behalf of its member growers, represents them in contract negotiations with processors, and maintains a marketing desk that will find outlets for members' apples as needed. The importance of the findings is that the members surveyed realize such efforts are being made on their behalf, and their perceptions differ from growers not in the organization.

Conclusions

The histories of the New York and Michigan apple industries indicate that state legislation with strong protections for growers to organize is essential for establishing viable bargaining cooperatives. It is also clear that the presence of a bargaining cooperative has enhanced the welfare of Michigan growers, especially MACMA members.

Comparisons of apple prices in Michigan, New York, Pennsylvania, and the United States as a whole from 1969 to 2002 showed that Michigan growers received higher prices for their apples through most of this period. The price benefit appears to be shrinking, however. Apple growers interviewed and surveyed for this study assign blame for falling U.S. apple prices to imported apples and apple concentrate from China.

The results of our mail survey indicate that MACMA members tend to be more satisfied than nonmembers. Holding a number of control variables constant, the survey analysis showed that MACMA members appear to reap substantial fringe benefits from their membership in the bargaining cooperative. These benefits include having input into contract terms and public policy that affects them as well as finding marketing assistance if needed.

Overall, we conclude that strong laws enabling the establishment of bargaining cooperatives, although not panaceas, help growers to maintain their operations in the face of structural change in the apple industry.

Appendix 10.A: Results from Grower Survey, Part I

Question	Agree (%)			Neutral (%)			Disagree (%)		
	*MAC	Non-MAC	NY	MAC	Non-MAC	NY	MAC	Non-MAC	NY
I have input into the price I received for my processed apple crop	26.3	7.0	8.1	13.5	5.6	4.7	60.3	87.3	87.2
I have input into terms of trade such as payment schedules for my processed apple crop	18.3	9.9	7.8	12.7	7.0	10.0	69.0	83.1	82.3
In general, apple growers in my state are at a disadvantage because the processing sector is too concentrated	38.8	38.6	73.9	30.3	41.4	17.8	31.0	20.0	8.3
In general, the apple-processing firms in my state have too much control over important aspects of my operation	54.6	43.7	63.2	26.2	29.6	22.1	19.1	26.7	14.7
I am generally satisfied with the marketing arrangements for my processed apple crop	20.4	15.5	11.6	16.2	23.9	18.5	63.4	60.6	69.8
We need a new and stronger federal law that requires processing firms to bargain with accredited grower bargaining cooperatives or associations	70.0	49.3	56.0	19.3	28.2	23.3	10.8	22.6	20.7
It is difficult to find useful apple market and production information that keeps me informed of industry developments	33.1	38.0	38.3	27.5	31.0	22.1	39.4	31.0	39.6

I have input into state government policies and programs that might affect my operation	29.6	18.3	24.7	26.8	25.4	22.6	43.6	56.3	52.8
I have input into federal government policies and programs that might affect my operation	21.8	16.9	13.7	25.4	23.9	22.2	52.8	59.1	64.1
If I need it, I can get assistance in finding a "home" or market outlet for my processed apple crop	51.4	17.1	13.2	14.8	24.3	13.7	33.8	58.6	73.1
Having a "home" or market outlet for my crop is more important than receiving a market price	12.0	12.8	19.4	13.4	21.4	18.5	74.7	65.7	62.1
We need more apple-processing firms in my state	64.8	50.7	83.4	23.2	36.6	13.7	12.0	12.6	3.0
In general, grower organizations that collectively bargain with processing firms for price, raise the price for all growers, even nonmembers	83.1	52.1	57.0	9.2	25.4	24.8	7.7	22.6	18.2
The price I receive for my processed apple crop is high enough to cover my production costs	9.9	7.0	10.4	6.3	14.1	11.2	83.8	78.8	78.5

* MAC = MACMA member; Non-MAC = Michigan grower but not a member of MACMA; NY = New York grower

Appendix 10.A: Results from Grower Survey, Part II

Do you process the apples you produce into a finished product?

Yes			No		
MACMA	Non-MAC	NY	MACMA	Non-MAC	NY
7.0%	5.6%	15.7%	93.0%	94.4%	84.3%

In the most recent year, what percentage of your farm's apples production was sold on the processed market?

	0–25	26–50	51–75	76–100
MACMA	12.0%	38.7%	19.7%	29.6%
Non-MAC	13.9%	26.4%	16.6%	43.1%
NY	27.5%	20.6%	22.3%	29.6%

In the most recent year, what percentage of your fresh market apples were packinghouse culls?

	0–10	11–25	26–50	51–100
MACMA	24.1%	45.8%	26.3%	3.8%
Non-MAC	41.4%	39.6%	17.3%	1.7%
NY	46.8%	38.1%	12.7%	2.4%

In the most recent year, what percentage of your apples for processing were purchased by processing firms located in your state?

	0–89	90–100
MACMA	12.6%	87.4%
Non-MAC	8.5%	91.5%
NY	47.6%	52.4%

Approximately how many processing firms do you have to choose from as potential buyers for your apple crop? (in state)

	1–3	4–6	7–14
MACMA	54.0%	34.5%	11.5%
Non-MAC	58.2%	32.8%	9.0%
NY	89.3%	9.8%	0.9%

Appendix 10.A
(continued)

Approximately how many processing firms do you have to choose from as potential buyers for your apple crop? (total)

	0–2	3–5	6–8	9–14
MACMA	28.0%	48.3%	16.2%	7.5%
Non-MAC	33.3%	47.7%	9.5%	9.5%
NY	38.7%	50.0%	9.1%	2.2%

In an average year (consider the last three years), how many processing firms do you deal with? (in state)

	0–3	4–9
MACMA	75.2%	24.8%
Non-MAC	75.4%	24.6%
NY	90.4%	9.6%

In an average year (consider the last three years), how many processing firms do you deal with? (total)

	0–3	4–12
MACMA	69.1%	30.9%
Non-MAC	69.2%	30.8%
NY	62.9%	37.1%

How often do you sell to a processing firm without an agreed-on price?

	Often	Sometimes	Seldom or never
MACMA	23.1%	25.2%	51.7%
Non-MAC	47.2%	27.8%	25.0%
NY	17.1%	30.3%	52.6%

In five years, will you or a member of your family be growing apples?

	Yes/ probably	Maybe/ don't know	No/ probably not
MACMA	66.4%	11.2%	22.4%
Non-MAC	69.5%	11.1%	19.5%
NY	68.1%	9.9%	22.0%

Appendix 10.A
(continued)

In an average year (consider the last three years), how many bushels of apples for processing do you produce?

	0–10,000	11,000–25,000	26,000–50,000	51,000–83,000	84,000–400,000
MACMA	33.3%	32.6%	24.0%	5.8%	4.3%
Non-MAC	30.4%	30.5%	24.6%	10.2%	4.3%
NY	35.1%	25.3%	21.6%	7.6%	10.4%

Gross processed apple sales volume

	MACMA	Non-MAC	NY
0–10,000	6.7%	9.7%	17.5%
10,001–20,000	15.6%	9.7%	13.9%
20,001–40,000	17.8%	25.0%	16.6%
40,001–60,000	17.0%	9.7%	8.5%
60,001–80,000	10.4%	8.3%	9.4%
80,001–100,000	10.4%	8.3%	5.8%
100,001–200,000	9.6%	23.6%	13.9%
200,001–300,000	5.9%	2.8%	6.7%
>300,000	6.7%	2.8%	7.6%

Do you own stock in an apple-processing cooperative?

	MACMA	Non-MAC	NY
Yes	58.7%	58.3%	20.9%
No	41.3%	41.7%	79.1%

If yes, do you own enough stock to meet all of your processed apple needs?

	MACMA	Non-MAC	NY
Yes	17.5%	23.6%	1.7%
No	42.0%	34.7%	19.0%
NA	40.6%	41.7%	79.2%

Note

1. The recent legislation is the *Family Farmer Cooperative Marketing Amendments Act of 2001*, HR 230, 107th Cong., 1st sess.

References

Babb, Emerson M., S. A. Belden, and C. R. Saathoff. 1969. An Analysis of Cooperative Bargaining in the Processing Tomato Industry. *American Journal of Agricultural Economics* 51:13–25.

Banker, David, and Janet Perry. 1999. More Farmers Contracting to Manage Risk. *Agricultural Outlook* (January–February): 6–7. Washington, DC: U.S. Department of Agriculture, Economic Research Service, Market and Trade Economics Division.

Brandow, George E. 1970. The Place of Bargaining in American Agriculture. In *Cooperative Bargaining: Selections from the Proceedings of the National Conferences of Agricultural Bargaining Cooperatives*. Service Report no. 113, 27–33. Washington, DC: U.S. Department of Agriculture, Agricultural Cooperative Service.

Bunje, Ralph B. 1980. *Cooperative Farm Bargaining and Price Negotiations*. Cooperative Information Report no. 26. Washington, DC: U.S. Department of Agriculture, Rural Business and Cooperative Services Agency.

Clodius, Robert L. 1957. Cooperatives in Changing Agriculture. *Journal of Farm Economics* 39 (5): 1271–1281.

Frederick, Donald A. 1990. Agricultural Bargaining Law: Policy in Flux. *Arkansas Law Review* 43 (3): 679–699.

Frederick, Donald A. 1993. Legal Rights of Producers to Collectively Negotiate. *William Mitchell Law Review* 19 (2): 433–455.

Garoyan, Leon, and Eric Thor. 1978. Observations on the Impact of Agricultural Bargaining Cooperatives. In *Agricultural Cooperatives and the Public Interest*. North Central Regional Research Publication no. 256, August, 135–148. Madison: University of Wisconsin, College of Agricultural and Life Sciences, Research Division.

Helmberger, Peter G., and Sidney Hoos. 1965. *Cooperative Bargaining in Agriculture: Grower-Processor Markets for Fruits and Vegetables*. University of California, Division of Agricultural Sciences.

Hoos, Sidney. 1962. Economic Possibilities and Limitations of Cooperative Bargaining Associations. In *Cooperative Bargaining: Selections from the Proceedings of the National Conferences of Agricultural Bargaining Cooperatives*. Service Report no. 113, 12–26. Washington, DC: U.S. Department of Agriculture, Agricultural Cooperative Service.

Iskow, Julie, and Richard Sexton. 1992. *Bargaining Associations in Grower-Processor Markets for Fruits and Vegetables*. ACS Research Report no. 104.

Washington, DC: U.S. Department of Agriculture, Rural Business and Cooperative Services Agency.

Lang, Mahlon G. 1978. Structure, Conduct, and Performance in Agricultural Product Markets Characterized by Collective Bargaining. In *Agricultural Cooperatives and the Public Interest*. North Central Research Publication no. 256, 118–234. Madison: University of Wisconsin, College of Agricultural and Life Sciences, Research Division.

Lang, Mahlon G. 1994. Where Bargaining Associations Fit. In *Agricultural Bargaining in a Competitive World*. Proceedings of the thirty-eighth National and Pacific Coast Bargaining Cooperative Conference, December 2–4, 1993, Portland, Oregon. Cooperative Services Report no. 42, 1–6. Washington, DC: U.S. Department of Rural Development Administration.

Levins, Richard A. 2001. *An Essay on Farm Income*. Staff Paper Series PO1–1. Saint Paul: University of Minnesota, Department of Applied Economics. Available at ⟨http://agecon.lib.umn.edu/⟩.

Looker, Dan. 1999. Cooperative Bargaining Pays Off: Apples and Asparagus Prices in Michigan Top the Nation, Thanks to a Strong Bargaining Co-op. *Successful Farming* (August): 28–31. Available at ⟨http://www.agriculture.com/sfonline/sf/1999/august/cooperative/index.html⟩.

Marcus, Gerald D., and Donald Frederick. 1994. *Farm Bargaining Cooperatives: Group Action, Greater Gain*. ACS Research Report no. 130. Washington, DC: U.S. Department of Agriculture, Rural Business and Cooperative Services Agency.

Morrison, John. 1995. Problems Facing New Bargaining Associations. In *Agricultural Bargaining in a Competitive World*. Highlights of the thirty-ninth National and Pacific Coast Bargaining Cooperative Conference, December 1–3, 1994, San Diego, California. Cooperative Services Report no. 47, 5–6. Washington, DC: U.S. Department of Agriculture, Rural Economic and Community Development Service.

Torgerson, Randall E. 1998. Cooperative Bargaining in the Next Millennium. *Agricultural Bargaining in a Competitive World*. Proceedings of the forty-second National and Pacific Coast Bargaining Cooperative Conference, December 4–6, 1997, Portland, Oregon. Service Report no. 55, 1–4. Washington, DC: U.S. Department of Agriculture, Rural Development Administration, Cooperative Services Agency.

USDA. 1977. Noncitrus Fruits and Nuts: Production, Use, and Value Estimates by State, 1969–74. Statistical Bulletin no. 593. Washington, DC: U.S. Department of Agriculture, Statistical Reporting Service.

USDA. 1980. Noncitrus Fruits and Nuts: Production, Use, and Value Estimates by State, 1974–78. Statistical Bulletin no. 647. Washington, DC: U.S. Department of Agriculture, Economics and Statistics Service.

USDA. 1985. Noncitrus Fruits and Nuts: Production, Use, and Value Estimates by State, 1978–82. Statistical Bulletin no. 718. Washington, DC: U.S. Department of Agriculture, Statistical Reporting Service.

USDA. 1990. Noncitrus Fruits and Nuts: Final Estimates, 1982–87. Statistical Bulletin no. 809. Washington, DC: U.S. Department of Agriculture, National Agricultural Statistics Service.

USDA. 1995. Noncitrus Fruits and Nuts: Final Estimates, 1987–92. Statistical Bulletin no. 900. Washington, DC: U.S. Department of Agriculture, Bureau of Agricultural Economics.

USDA. 1998. Noncitrus Fruits and Nuts: Final Estimates, 1992–97. Statistical Bulletin no. 950. Washington, DC: U.S. Department of Agriculture, National Agricultural Statistics Service.

USDA. 1999. Noncitrus Fruits and Nuts: 1998 Summary. Fr Nt 1–3 (99). Washington, DC: U.S. Department of Agriculture, National Agricultural Statistics Service. Available at ⟨http://usda.mannlib.cornell.edu/reports/nassr/fruit/pnf-bb/⟩.

USDA. 2000. Noncitrus Fruits and Nuts: 1999 Summary. Fr Nt 1–3 (00)a. Washington, DC: U.S. Department of Agriculture, National Agricultural Statistics Service. Available at ⟨http://usda.mannlib.cornell.edu/reports/nassr/fruit/pnf-bb/⟩.

USDA. 2001. Noncitrus Fruits and Nuts: 2000 Summary. Fr Nt 1–3 (01). Washington, DC: U.S. Department of Agriculture, National Agricultural Statistics Service. Available at ⟨http://usda.mannlib.cornell.edu/reports/nassr/fruit/pnf-bb/⟩.

USDA. 2002. Noncitrus Fruits and Nuts: 2001 Summary. Fr Nt 1–3 (02). Washington, DC: U.S. Department of Agriculture, National Agricultural Statistics Service. Available at ⟨http://usda.mannlib.cornell.edu/reports/nassr/fruit/pnf-bb/⟩.

USDA-ERS. 1996. *Farmers' Use of Marketing and Production Contracts.* Agricultural Economic Report no. 747. Washington, DC: U.S. Department of Agriculture, Economic Research Service, Rural Economy Division, Farm Business Economics Branch.

USDA-RBS. 2001. *Directory of Farmer Cooperatives.* RBS Service Report no. 22. Washington, DC: U.S. Department of Agriculture, Rural Business-Cooperative Service. Available at ⟨http://www.rurdev.usda.gov/rbs/pub/sr22.pdf⟩.

U.S. Department of Labor. 2002. Consumer Price Index—All Urban Consumers, All Items. Available at ⟨ftp://ftp.bls.gov/pub/special.requests/cpi/cpiai.txt⟩ (accessed August 8, 2002).

11

Sustaining the Middle: The Roles of Sustainable Agriculture and Agri-Environmental Payments

Sandra S. Batie

U.S. agriculture has been heavily influenced by seven decades of governmental programs that have had significant influence on farm size and vitality. Yet there has been scant direct policy attention to the size or ownership of farms—that is, the issue of the "structure" of U.S. farms. This statement remains true despite occasional attempts to elevate the political relevancy of the structure issue. Examples include the A *Time to Choose* project (USDA 1981) and the recent lobbying by numerous nongovernmental organizations in an attempt to influence the design of the 2002 Farm Bill with respect to structure issues.

In contrast to this neglect of farm structure, agricultural policy has increasingly focused on agri-environmental issues. Over the last several decades, agricultural policy has incrementally shifted its focus away from resource development for agricultural production toward agri-environmental protection (Batie 2001; Cox 2001). Currently, the agri-environmental policy debate is centered around the wisdom and design of green payments—government payments to producers for the provision of agri-environmental services. Agri-environmental services can include the provision of high-quality air and water, wildlife habitat, landscape amenities, flood control, nutrient recycling, and even carbon sinks. Agricultural lands can also provide the services associated with hunting, agritourism, and agri-entertainment as well as those of regional identity, heritage values, rural vitality, and countryside ambience.

The European term for an agricultural sector that provides many agri-environmental services is "multifunctional agriculture." Many European nations have supplied payments to farmers for the provision of these services (Aldington 1998; OECD 2001). In Europe, the definition of green payments frequently includes regional, rural development, cultural, or structural goals. This definition contrasts with that used in the United States, where the term green payments usually only refers to payments to producers for agri-environmental services.

Several forces have combined to enhance U.S. public interest in green payments for agri-environmental services. First, agri-environmental services are increasingly valued as incomes rise, as they have in the United States (Batie 2003; Hellerstein et al. 2002; Schweikhardt and Browne 2001). Furthermore, the United States is increasingly an urban society. When rising incomes are combined with urban values, the result is that the public demands more agri-environmental services from agriculture (Batie 2003; Schweikhardt and Browne 2001). Finally, as nations agree to pursue free trade policies, such as those administered by the World Trade Organization, the nations must reduce or eliminate trade-distorting direct payments to producers that are based on crop production. Depending on their implementation details, green payments may be viewed as a nontrade distorting means for supporting the agricultural sector of nations, and thus should not violate trade agreements (Ervin 1999). This fact has heightened interest in expanding the use of green payments, even by traditional audiences not usually concerned with agri-environmental services.

Now, as evidence of the increasingly bipolar nature of the structure of agriculture becomes better understood, there is interest in exploring how agricultural policy might be remodeled to better support the farms in the middle—roughly defined as those farms that have annual sales between $100,000 and $250,000. The most cost-effective manner of supporting these farms in the middle would be with a direct lump-sum payment from the taxpayers to the owner-operators of the midsize farms. But at the national level, the political authorization of such lump-sum transfers are unlikely.

An interesting question, then, is whether the public support of green payments could be capitalized on to support midsize farms as well. That is, could green payments be used to achieve new potential policy objectives—such as the support of midsize farms? It is known, for example, that small and midsize farms—those with less than $250,000 in annual sales—account for 91 percent of the farms, 68 percent of the land assets, and 33 percent of the total production in U.S. agriculture (Hoppe 2001). Therefore, programs that effectively influence the management of some two-thirds of U.S. farmland could play a major role in the provision of agri-environmental services. At the same time, small and midsize farms have significantly smaller average operator household income than do farms with sales greater than $250,000 per year, and the bulk of unprofitable farms are those making less than $250,000 per year (ibid.). Green payments might be a welcome income enhancement for these farmers.

This chapter will explore what we know about the potential use of green payments to pursue the dual objectives of the provision of agri-environmental services and the income support of the owner-operators of midsize farms. There are many potential agri-environmental services; this chapter will focus on the amelioration of nonpoint pollution problems. It assumes that the public demand for such support of agri-environmental services is strong, as the evidence suggests (Batie 2003; Esseks and Kraft 2002; Kline and Wichelns 1996). It also assumes, albeit with less evidence, that the public demand for the support of midsize farms is strong. Thus, rather than investigating the desirability of this objective, this chapter concentrates on issues associated with using green payments to support the income of midsize farms.

In addition to addressing policy design challenges, this chapter asks the question: To what extent will green payments for agri-environmental services increase the income of midsize U.S. farms? What becomes apparent with this investigation is that green payments are an expensive and blunt instrument for supporting the income of midsize farms; alternative options for the support of midsize farms are presented in the conclusion section.

This chapter starts with a brief historical perspective to provide a context about green payment policy, and then proceeds to examine the design issues of green payment programs.

U.S. Green Payments: A Historical Perspective

The most recent federal agricultural green payment program is the fledging Conservation Security Program (CSP), an agri-environmental program in the 2002 Farm Bill—the Farm Security and Rural Investment Act of 2002. The CSP supplements other federal agricultural programs and authorizes payments to producers for undertaking a variety

Box 11.1
The Conservation Security Program

> The CSP is a 2002 Farm Bill initiative that provides green payments to producers who voluntarily solve certain resource and agri-environmental problems associated with their farm operation. It is a program focused on "working lands," as opposed to the Conservation Reserve Program that idles cropland. Initially, the program was authorized as a national program, available to all producers in all fifty states. The program, however, was capped at $41 million for its first year and limited to eighteen watersheds throughout the country. In 2005, the program was offered nationwide, and the number of participating watersheds was significantly increased. There are minimum conservation requirements for each of three levels or tiers of participation. Tier I participation requires the producer to address certain water- and soil-quality problems on the part of their farm or ranch, whereas Tier II requires the addressing of such problems on the entire farm. For example, Tiers I and II for pasture and rangeland require the following of a grazing management plan that provides a forage-animal balance, proper livestock distribution, and managing livestock access to watercourses. Tier III participants must address all the resource concerns on the entire farm.

of agri-environmentally oriented changes on their farms (for a brief description of the CSP, see box 11.1).

The CSP, as a green payments program, has many antecedents (Helms 2003). Indeed, producers have received payments for conservation efforts since the beginning of farm programs in 1936.[1] But these early conservation programs were almost exclusively focused on protecting on-farm fertility or with idling croplands.

Starting in the 1970s, NGOs began promoting "red ticket" and "green ticket" approaches to government agricultural programs (Helms 2003). Red ticket approaches were those that punished farmers who did not follow appropriate conserving practices, whereas green tickets were payments that rewarded producers who followed appropriate conserving practices. By the mid-1970s, the Soil and Water Resources Conservation Act of 1977 directed the secretary of agriculture to develop policies and programs that would conserve the nation's soil and water resources (ibid.). Some of the resulting programs followed the red ticket approach. An example is the "cross-compliance" requirement, which denied commodity and other program benefits to producers who did not meet specified erosion standards (Claassen et al. 2004).

Many subsequent programs, though, were green ticket approaches that paid producers for desired actions. For example, the 1985 Farm Bill authorized the largest conservation land retirement program—the Conservation Reserve Program (CRP). The CRP pays producers to idle all or some of their farm or ranch for conservation purposes (Hummon and Casey 2004). While technically a green payment program, the CRP continued a long-standing federal policy of paying for idling land—thus acting as a supply control as well as a conservation measure. By idling lands, budgetary outlays for commodity payments were reduced. After 1990, the CRP redirected its eligibility criteria from that of highly erodible lands toward more water-quality and wildlife habitat goals (Heimlich 2003).

The CRP is a relatively expensive way of obtaining agri-environmental services (USDA-ERS 1997; Claassen et al. 2001). Nevertheless, by 2000, about 85 percent of the USDA conservation budget was spent on land retirement assistance in the CRP, whereas only 15 percent was for land treatment on working acres (SWCS 2000). This 15 percent included other green payment programs such as the Environmental Quality Incentives Program (EQIP), the Wetlands Reserve Program (WRP), and the Wildlife Habitat Incentive Program (WHIP) (for more discussion of these examples, see box 11.2).

While green payment programs have been present in U.S. agricultural policies for the last three decades, it is clear that politicians have not given the same importance to green payment objectives that they do to other Farm Bill objectives. This fact is clearly reflected in the distinctly different funding differences between the traditional commodity program payments and those directed toward conservation. Historically, with the exception of the CRP, green payment programs have not transferred enough dollars to farm owners to have any impact on farm incomes (Horan et al. 1999). Recent increases in the conservation funding following the 2002 Farm Bill (the 2002 Farm Security and Rural Investment Act), however, have been impressive. The act authorizes increases in conservation funding to levels that by 2007 will be double those of the last decade, with two-thirds of the funds going to programs emphasizing

Box 11.2
Examples of Existing Green Payments Programs

The CRP was established in the 1985 Farm Bill and is a voluntary cropland retirement program with a maximum enrollment of 36.4 million acres. The current enrollment is approximately 34 million acres. The program provides farmers payments for ten to fifteen years for land retired from crops and placed into permanent cover. Parcels are selected based on the magnitude of the likely agri-environmental services relative to the rental payment. Agri-environmental services include habitat improvements, water-quality impacts, soil productivity gains, quality improvements, and carbon sequestrian.

EQIP was established in the 1996 Farm Bill and consolidates other programs to better target them to agri-environmental concerns. The objective of EQIP is to encourage producers, including livestock producers, to adopt practices that reduce agri-environmental problems. In 2004, EQIP allocated $908 million to the states.

The WRP was authorized by the 1990 Farm Bill, and provides an easement payment and covers wetland restoration costs for land permanently converted back to wetlands. The enrollment goal cap was recently increased to 2,275,000 acres, with the goal of enrolling 250,000 acres per year until the maximum program acres cap is achieved.

WHIP was authorized by the 1996 Farm Bill and encourages the creation of wildlife habitats that support wildlife populations. Currently, WHIP has more than 1.6 million acres enrolled.

For more information, see Claassen (2007).

conservation on working lands (Claassen 2007). Partly as a result of the new funding, attention has turned to the alternative design of green payment programs.

The Design of a Green Payments Program

It is clear that in principle, it is possible to design a green payments program in support of midsize farms—as the existence of current green payment programs attest. National data are now available that include, by region, the size, location, and profitability of farm and ranch enterprises.[2] Small and midsize farms (i.e., with annual sales of less than $250,000) account for only about a third of the value of agricultural production, but have large shares of particular commodities such as 62 percent for hay, 54 percent for tobacco, 40 percent for soybeans, 47 percent for wheat, 47 percent for corn, and 40 percent for beef (Hoppe 2001).

Our knowledge—with respect to the general location, nature, and magnitude of many agri-environmental problems—has also greatly expanded. It is clear that actual and potential agri-environmental problems differ by source and impact, and these problems are unevenly distributed throughout the nation (Heimlich 1994; Claassen et al. 2001, 2004).

These sets of data—on midsize farms and agri-environmental problems—can be overlapped so as to analyze and design alternative green payment programs for the dual objectives of addressing nonpoint pollution and supporting midsize farm income. There are many green payment program alternatives, however; each will have different trade-offs. That is, each program choice—of who will be eligible to receive payments as well as what situation constitutes eligibility—will result in different impacts by the size of the farm or ranch, the region, the agri-environmental outcome, and the program cost.

Any green payment program must resolve several issues. The design challenge can be summarized as: "How much is paid to whom for taking what action on what land?" (Claassen et al. 2001). These issues include the following (Batie and Horan 2001; Heimlich 1994; Claassen and Horan 2000):

• What is the objective of the program?
• Who should be paid?
• How much should farmers and ranchers be paid?
• What should farmers and ranchers be paid to do?

The answers to these questions determine the cost-effectiveness of the green payments program, the administrative and implementation costs, and the resulting trade-offs.

What Is the Objective?

The choice of objective(s) is crucial. A program that is directed toward only enhancing agri-environmental services or only supplementing midsize farm income will be more cost-effective than one with dual objectives (Batie and Horan 2001). The reason for this lack of cost-effectiveness is that there is not a one-to-one relationship between midsize farms and profitability, nor between either size and profitability and any particular agri-environmental service. For example, research has shown that directing green payments to farms with limited income—many of which would be small and midsize farms—would exclude large amounts of acreage with agri-environmental problems (Claassen et al. 2001). Similar problems would exist if payments were directed only by the size of the farm, regardless of its financial status.

Also, directing payments to a single agri-environmental problem would not necessarily address problems stemming from another agri-environmental problem (ibid.). For instance, reducing nitrogen runoff into surface waters may lead to increasing nitrogen leaching into groundwater.

Furthermore, each of these different policy alternatives would lead to a different "geography of payments." For example, a green payments program will have a different geographic impact if it focuses on, say, water-quality damages rather than the incidence of rainfall erosion. Water-quality damage objectives would result in those farms in the Great Lakes drainage basin, or on the Pacific, Atlantic, or Gulf coastal plains, receiving more green payments than would inland farms (ibid.). Targeting the incidence of rainfall erosion would include more inland farms.

Because of the multiple possible targets, one possible approach is to direct green programs to water-quality problems via the use of a benefits index. Such an index would be similar to that used for the CRP (for a discussion of the CRP benefits index, see box 11.3). The index would have several weighted components of various components of water quality as well as some measure of the farm size or financial situation. The objective, then, of such an index-based green payments program would be to target payments to those farms where the benefits index is high.

Who Should Be Paid?

The issue of who should be paid may not be the same as which land should be included in the program. For example, a green payments program could be designed in such a manner that all producers operating midsize farms and ranchers are eligible, and if they participate, are paid uniformly for using certain agri-environmental-enhancing practices. This

Box 11.3
Environmental Benefits Index for the CRP

An environmental benefits index is used to prioritize and rank offers for land to be enrolled into the CRP. The index is the sum of six ranked environmental factors plus a cost factor. The following are the factors that comprise the index:

N1 = Wildlife habitat benefits
N2 = Water-quality benefits from reduced water erosion, runoff, and leaching
N3 = On-farm benefits of reduced wind or water erosion
N4 = Long-term benefits of certain practices that will likely extend beyond the contract period
N5 = Air-quality benefits from reduced wind erosion
N6 = Benefits from enrollment in conservation priority areas when the offer significantly contributes to the priority area concern
N7 = Government cost of the contract

wide-scale eligibility based on producers' actions is the concept underlying the CSP. Still, the implementation of the CSP has been slowed and drastically reduced in scope from the original authorization, in part because of major budgetary exposure and implementation challenges.

Wide-scale eligibility for a green payments program should lower many administration costs and appears, to many observers, to be more equitable. Yet even such a straightforward design has its drawbacks. First, with respect to the agri-environmental objectives, uniform payments may reduce the overall cost-effectiveness because they encourage farmers with few agri-environmental impacts or low costs of compliance to participate in large numbers, while farmers with large agri-environmental impacts and/or high costs of compliance would not have payment incentives to participate. Paying for a certain agricultural practice instead of an outcome also limits producer flexibility to select the least-cost agri-environmental management technique. Because they are not tailored to the producer's situation, uniform practices do not always result in agri-environmental improvements (Batie and Horan 2001). Then, if the further complication of support of midsize farm income is added, there will be the additional slippage of meeting both the agri-environmental and income objectives, because there is not a direct correspondence between farm size and agri-environmental services.

An alternative green program design is to limit eligibility. One method of doing this is to use spatial targeting, so that eligibility is predetermined by the location of the agricultural land. For example, a midsize farm or ranch operator may be deemed eligible to receive green payments only if the farm or ranch is located in a certain watershed where the protection of water quality from agricultural pollution is an objective. While such an approach can reduce program costs, midsize farms and ranches not in the selected watersheds would not be deemed eligible, and therefore would not be assisted by green payments. This limited eligibility design is the same as the current implementation of the CSP.

There are other design possibilities. For example, there could be financial incentives within a green payments program for farmers and ranchers to cooperate, and manage their farms and ranches in a coordinated manner so as to provide a contiguous block of land for such agri-environmental services as the provision of wildlife habitat. If a major objective of the program were the support of only midsize farms, though, problems would emerge if a large farm or ranch was a particularly crucial component within the contiguous block of land.

How Much Should They Be Paid?
In a voluntary green payments program, participation will be minimal unless producers receive enough payments to offset any opportunity costs they incurred with respect to the provision of the agri-environmental services. It is difficult to determine the level of payments that will both motivate changes that would not otherwise have taken place, and will neither over- nor undercompensate the producer. On the other hand, if overcompensating the producer is the objective of the green payments so as to increase the income of midsize farms, then not only can the budgetary exposure be great, it may be that the payments will be viewed as trade distorting. Furthermore, unless there is "means testing"—that is, a determination of the true level of financial need of the operator, and basing eligibility on financial need—then many financially healthy producers may receive payments.

What Should Farmers and Ranchers Be Paid to Do?
Green payments can be based on either agri-environmental performance (outcomes) such as achieving improved water quality or the owner-

operator taking some action such as using a particular environmentally friendly practice (Hrubovcak et al. 1999). In either case, the selection of the appropriate baseline from which to measure changes will be critical in affecting both program participation and the level of payments. If the baseline is too stringent, then few producers will participate. If it is too lax, the program will be paying farmers and ranchers for actions they should already be doing to meet current agri-environmental regulations. Payments to these producers implicitly penalize those who have already taken steps to improve agri-environmental services (Batie and Horan 2001). Producers sometimes refer to the latter problem as "rewarding the bad actor." In the worst case, too lax a baseline could create an opportunity for "moral hazard"—that is, a producer creating agri-environmental problems in order to be compensated for ameliorating them (ibid.).

Producers could be paid on the basis of the practices adopted rather than on any geographic characteristic or agri-environmental outcomes. With this design, a key question for a program with the dual objectives of agri-environmental protection and midsize farm income support is: What set of practices could be funded that would score high on both agri-environmental outcomes and midsize farm income support?

Regardless of whether the program pays midsize farm owner-operators to adopt certain practices (i.e., filter strips) or pays farmers to achieve certain water-quality standards, there are many problems to resolve. The reason for this situation is that the relationships among management practices on specific farms and the ultimate effects on agri-environmental services are not completely understood (Claassen and Horan 2000). For example, there is difficulty in observing and measuring impacts of changing farm practices because of the diffuse nature of non-point pollution and the unpredictability of natural events. This situation is also complicated by the diversity of agriculture—practices, crops, and topography all vary widely among producers (ibid.).

There are several ways that midsize farm income could be impacted by green payments—both positively and negatively. For instance, Roger Claassen and Richard Horan (2000) note that farm income could be affected by green program design via payment size limitations and distribution (e.g., more and larger payments to midsize farms), changes in direct farm costs resulting from changes in production practices and enterprise mix, cropping patterns, and crop yields, and changes in market prices as a result of the program.

The Economic Research Service Studies

Two recent national studies conducted by the Economic Research Service (ERS) of the USDA illustrate some of the different "geography of payments" that might result from different green payment programs (Claassen et al. 2001, 2004). Rather than only looking at those midsize farms with annual sales of between $100,000 and $250,000, these studies instead aggregated these farms together with smaller-size farms with sales of less than $100,000 per year. Still, the results are quite revealing with respect to green payments program design. In the 2001 study, ERS researchers examined the relationship that small, midsize (less than $250,000 in annual sales), and "moderately unprofitable" farms have to acreage that has agri-environmental problems.[3] The agri-environmental problems explored for these two types of farms were, by region, rainfall erosion acreage (where the rainfall erosion rates were higher than soil tolerance "T" levels), wind erosion acreage (where the wind erosion rates on nonhighly erodible cropland were greater than soil tolerance "T" levels), and nitrogen runoff acreage (where the nitrogen runoff to surface water was estimated to exceed one thousand kilograms per square kilometer per year). Nitrogen runoff is a major contributor to estuarine hypoxia problems, such as the "dead zone" in the Gulf of Mexico (Goolsby et al. 1999).

Agri-Environmental Problems as the Green Payments Objective

The researchers had the following conclusions with regard to directing the program at agri-environmental problems. First, if a green payments program were to be directed at these three problems (wind erosion, rainfall erosion, and nitrogen runoff), approximately 70 percent of the small, midsize, and moderately unprofitable farms would be eligible to participate.[4] Yet the percentages of eligible farms vary widely by region. For example, more than 95 percent of the small, midsize, and moderately unprofitable farms in the Heartland region would qualify, compared to only 34 percent of the small and midsize farms and less than 40 percent

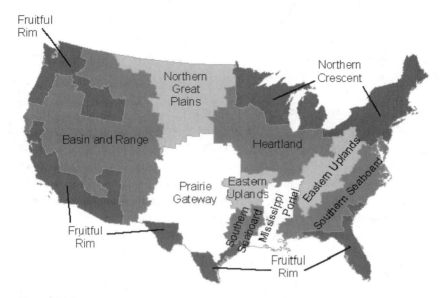

Figure 11.1
Farm Resource Regions
Source: Available at 〈http://www.ers.usda.gov/Emphases/Harmony/issues/
resourceregions/resourceregions.html〉.

of the moderately unprofitable farms in the Eastern Uplands area. These
regions' boundaries are shown in figure 11.1.

There is considerable variation by region if different agri-
environmental objectives were selected for the green payments program.
The Heartland region, for example, has considerable acreage with rainfall erosion and sediment problems, and would receive the large share
of green payments, if these payments were based on potential sediment
damage. In contrast, if the program focused on nitrogen damage, green
payments would be directed to the Southern Seaboard region (Claassen
et al. 2001).

Of course, if the green payments program were expanded to include
more agri-environmental problems, then more farms in more regions
would qualify. This expanded program scope, however, would mean
that more program funds would be directed at farms that do not fall in
the category of moderately unprofitable (ibid.). Indeed, if the budget for
green payments were large enough, and if the support of a certain size

farm were not an objective, then all farmers and ranchers could be eligible to receive payments for either individual practices, whole farm conservation plan implementation, or even several cooperating farms' agri-environmental improvements.

Midsize Farms as a Green Payments Target

An alternative approach, less costly than wide-scale eligibility, would be to direct green payments not to agri-environmental objectives but rather to midsize or moderately unprofitable farms, or both. This approach results in trade-offs with respect to the agri-environmental services obtained. The ERS researchers concluded that if a green payments program were directed solely to farms in need of income support, or only at small and midsize farms, then less than half of all rainfall erosion, wind erosion, and nitrogen runoff acres would be impacted (Claassen et al. 2001).

In this case, the overlap of small and midsize farms with moderately unprofitable ones is modest. "While small [and midsize] farms contain over 40 percent of the acres with rainfall erosion problems and nitrogen runoff problems, . . . only about 30 percent of these acres are likely to be located on moderately unprofitable farms" (ibid., 29). The researchers conclude: "More generally, targeting any group defined by gross sales or source of household income ('farm' vs. 'nonfarm') is unlikely to capture a majority of agri-environmental problems, unless the criteria are very broadly defined. No single group defined by the ERS farm typology accounts for more than 25 percent of any of our agri-environmental indicator acreages" (ibid., 29).

Green Payments as Substitutes for Commodity Payments

Because budgetary limitations can seriously limit green payments programs, an interesting question is whether green payments would be effective substitutes for existing commodity payments. In the second ERS study (Claassen et al. 2004)—which examined conservation compliance as a policy vehicle—ERS researchers, using data from the Agricultural Resource Management Survey, identified the geographic distribution of farms receiving federal program payments. The programs included were disaster payments, commodity program payments, and conserva-

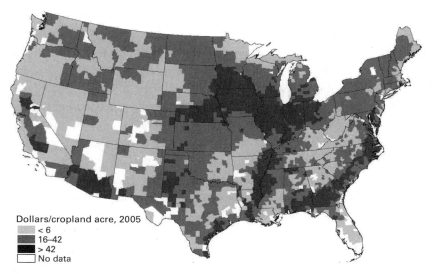

Dollars/cropland acre, 2005
- < 6
- 16–42
- > 42
- No data

Figure 11.2
Spatial Distribution of Commodity Program Payments
Source: Farm Service Agency.

tion payments from the CRP, WRP, and EQIP. Farms receiving some type of government payment accounted for 86 percent of U.S. cropland.

Figure 11.2 shows the spatial distribution of commodity program payments in 1998. The cropland receiving commodity payments is concentrated in several areas (e.g., the Heartland region, with 22 percent of cropland; the Prairie Gateway, with 17 percent of cropland; or the Northern Great Plains, with 17 percent of cropland). Many fewer commodity program dollars per acre are garnered in the Northern Crescent (9 percent of cropland), the Eastern Uplands (6 percent of cropland), or the Basin and Range (4 percent of cropland) regions. The Eastern Uplands and the Northern Crescent, for example, have many small and midsize farms. But these regions do not receive as many commodity program payments as do other regions.

The reason for such disparity is that many agricultural subsectors are not influenced heavily by commodity programs. These subsectors include vegetables, fruits, and livestock. Converting commodity program payments into green payments without expanding coverage to include these subsectors would exclude many small and midsize farms. Such an approach would also exclude many agri-environmental problems. In addition, if the green payments, as substitutes for commodity payments, were redirected to operators of only small and midsize farms, there would be a substantial redistribution of payments between regions and operators—a daunting task to accomplish politically.

On the other hand, green payments could replace commodity program payments in some regions and have significant influence on some agri-environmental outcomes.[5] Consider an agri-environmental problem such as hypoxia in the Gulf of Mexico. Seventy-two percent of U.S. cropland is located in the Heartland, Eastern Uplands, Mississippi Portal, Northern Great Plains, and Prairie Gateway regions. These are approximately the same regions that contribute to nitrogen loadings into the Gulf of Mexico and the hypoxia problem, as shown in figure 11.3. As would be expected, these regions are also where key commodity program payments—Market Loss Assistance, Loan Deficiency Payments, and Production Flexibility Contracts—are concentrated. This region also contains about 60 percent of the low-sales (with farming as the operator's major occupation) small and midsize farms, and 65 percent of the high-

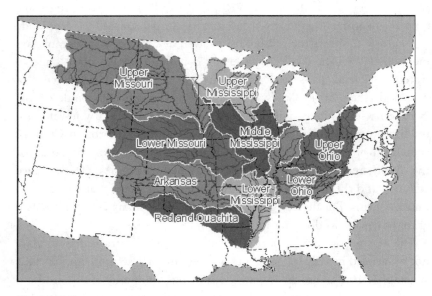

Figure 11.3
Drainage Subbasins Contributing to Hypoxia in the Gulf of Mexico
Source: Available at ⟨http://co.water.usgs.gov/hypoxia/html/graphics2.html⟩.

sales (with farming as the operator's major occupation) small and mid-size farms, and over 30 percent of the resource-limited small and midsize farms.

The ERS researchers (Claassen et al. 2004) found that the farms with the highest nitrogen runoff potential tend to participate heavily in farm programs and receive larger than average per acre payments (see figure 11.4). The abolition of commodity program payments might make it politically feasible to fund a green payments program where all farmers in these regions were eligible to receive green payments.[6] The payments would serve as incentives for producers to reduce nitrogen runoff by reducing applications of fertilizers and nutrients for either using filter strips or restoring wetlands along rivers and streams (Ribaudo et al. 2001).[7] Because two-thirds of the nitrogen in the Mississippi River comes from the use of agricultural fertilizers and manure on agricultural lands, these payments would be expected to result in significant reductions in nutrient runoff (ibid.).

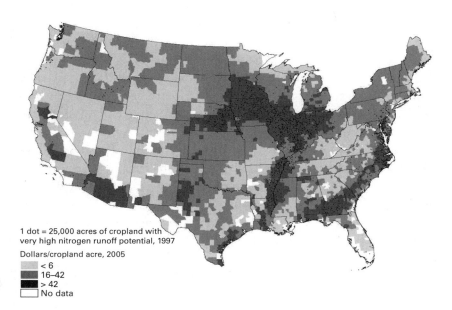

1 dot = 25,000 acres of cropland with very high nitrogen runoff potential, 1997

Dollars/cropland acre, 2005
- < 6
- 16–42
- > 42
- No data

Figure 11.4
Distribution of Commodity Program Payments and Very High Nitrogen Runoff Potential, 1997
Source: Farm Service Agency and the ERS.

This conclusion assumes that the green payments to reduce nitrogen runoff more than covered the costs of changed practices. If payments were only given to operators of small and midsize farms, however, there would be considerably less impact on total runoff reductions, since these same regions have the bulk of the large, very large, and nonfamily farms.

The fact that certain regions have a better correspondence of midsize farms to environmental problems does suggest that it would be valuable to examine the devolution of green payments to states so that they could better tailor programs to their region's situation. Such devolution could involve block grants to states accompanied by flexible criteria for achieving agri-enviromental and/or midsize farm income support objectives (Offutt et al. 2004). Indeed, one method of distributing federal funds would be via a "means test" so as to make midsize farms eligible for assistance (ibid.).

Trading Green Payments Credits

Because an extensive green payments program is expensive, and because taking funds from the commodity program is politically difficult, there is reason to identify green payments programs that generate their own revenues. One possibility would be to establish a national or regional trading of green credits. The concept is fairly simple: the government creates a supply of green credits by establishing a baseline of producer actions or performance outcomes from which to measure changes, and then gives green credits to producers for agri-environmental improvements beyond a minimum level of agri-environmental stewardship. For trading programs to be successful, governments need to create demand for the purchase of green credits as well as a supply. One method to create demand would be the rigorous regulation of point and nonpoint pollution sources. Mandatory caps would need to be placed on pollution discharges. At the same time, these sources, within these regulations, would need to be provided the option of purchasing credits in lieu of accomplishing all required pollution prevention from their own source.

While this design is conceptually simple, it is quite difficult to accomplish in practice. Practical challenges loom large, and include the lack of appropriate regulation, the existence of regulation that serves as a barrier to trade, the difficulties of determining when a green credit has been generated, and the difficulties of establishing a robust market in credits as well as auditing, monitoring, and enforcement problems (King and Kuch 2003; Stephenson et al. 2005). These problems would be made all the more challenging if the dual objective of supporting the income of small and midsize farm owner-operators were included as a trading objective.

Conclusion

The support of midsize farms as a policy objective would be most cost-effectively pursued with a lump-sum transfer of dollars from taxpayers to eligible operators. For a variety of reasons, such lump-sum transfers are not usually the chosen policy vehicle. There has been interest in examining, therefore, whether support for midsize farms could be obtained via a green payments program. But coupling the support of small farm operators with other objectives—such as the provision of agri-environmental services—necessarily results in trade-offs. As studies have shown, unless the budget allocated to a green payments program is extensive, then the pursuit of one objective (i.e., the support of midsize farm owner-operators' incomes) comes at the expense of another (i.e., agri-environmental services).

What becomes clear from the examination of the evidence to date is that the use of national green payments to support the income of midsize farms is a blunt instrument involving significant slippage on the achievement of objectives. On the other hand, as the hypoxia problem has illustrated, a state-tailored green payments program that also targets midsize farm income is worthy of a more careful exploration.

Nevertheless, it will probably be more cost-effective to pursue the support of midsize farms through other policy vehicles. The success of many of these midsize farms depends on appropriate market structures that provide them access to consumers and allows them the opportunity to take advantage of emerging value chains. To assist midsize farms, a more cost-effective public investment would be for government programs to improve midsize farm owner-operators' skills, knowledge, and management as well as develop and improve access to consumer markets. (See chapter 1, this volume.)

Notes

Thanks are due to Janet Larson, Michigan State University graduate research assistant, for her help in developing this chapter.

1. For a discussion of the history of conservation programs, see Batie (1985). The 1936 programs provided a constitutional approach for paying farmers to idle land for "conservation" objectives, but these were the same lands contributing to crop surpluses and hence low prices. So the dual objectives of supply control and conservation were pursued with one policy vehicle: conservation payments. These dual objectives have continued to influence agricultural policy throughout its history.

2. For a description of these data sources and how they can be used to analyze alternative policy designs, see the appendixes in Claassen et al. (2001, 2004). For a good overview of the Agricultural Resource Management Study, see Reinsel (1998). This study gathers data to show a detailed picture of the economics of agricultural production, and is a primary source of information about resource use, costs of production, the environment, and the structure and economic well-being of farm businesses and households.

3. Moderately unprofitable farms are those "where the full (economic) costs of production exceed total revenue by up to 50 percent. These farms are not financially viable..., but are more likely than higher cost firms to become so through government support payments.... If a policy goal is to keep farmers in farming, income support may be most helpful if directed toward moderately unprofitable farms" (Claassen et al. 2001, 30). Not all the moderately unprofitable farms are small or midsize ones.

4. Green payments could also be used to complement the implementation of the Section 319 Nonpoint Pollution Source Program of the Clean Water Act that is administered by the Environmental Protection Agency. That is, payments could go to watersheds already identified as part of the total maximum daily load regulations as needing to reduce nonpoint pollution from agriculture.

5. Instead of a green payments program, the commodity payments could be contingent on an agri-environmental compliance requirement. This red ticket approach is discussed in Claassen et al. (2004).

6. The Southern Seaboard area also has hypoxia problems, has a significant number of small and midsize farms, and receives high amounts of commodity payments per acre. This region could be included in a green payments program designed to reduce hypoxia conditions in marine estuaries.

7. A recent survey article by Michael Dosskey (2001) concluded that on average, filter strips remove 50 to 90 percent of nitrogen and phosphorus from runoff.

References

Aldington, Tim J. 1998. *Multifunctional Agriculture: A Brief Review from Developed and Developing Country Perspectives.* Internal Document no. 29, November. Rome: Food and Agriculture Organization of the United Nations, Agriculture Department.

Batie, Sandra S. 1985. Soil Conservation in the 1980s: A Historical Perspective. *Agricultural History* 59 (2): 107–123.

Batie, Sandra S. 2001. *Public Programs and Conservation on Private Lands.* Paper presented at Private Lands, Public Benefits: A Policy Summit on Working Lands Conservation, National Governors' Association, Washington, DC.

Batie, Sandra S. 2003. The Multifunctional Attributes of Northeastern Agriculture: A Research Agenda. *Agricultural and Resource Economics* Review 32 (1): 1–8.

Batie, Sandra S., and Richard Horan. 2001. Green Payments Policy. In *The 2002 Farm Bill: Policy Options and Consequences*, ed. Joe Outlaw and Edward G. Smith, 163–168. Chicago: National Public Policy Education Committee, Agriculture and Food Policy Center, Farm Foundation.

Claassen, Roger. 2007. Emphasis Shifts in U.S. Conservation Policy. *Amber Waves* (May): 28–34. Washington, DC: U.S. Department of Agriculture, Economic Research Service.

Claassen, Roger, Vince Breneman, Shawn Bucholtiz, Andrea Cattaneo, Robert Johansson, and Mitch Morehart. 2004. *Agri-Environmental Compliance in U.S. Agricultural Policy: Past Performance and Future Potential.* USDA-ERS Agricultural Economic Report no. 832. Washington, DC: Economic Research Service.

Claassen, Roger, LeRoy Hansen, Mark Peters, Vince Breneman, Marca Weinberg, Andrea Cattaneo, Peter Feather, Dwight Gadsby, Daniel Hellerstein, Jeff Hopkins, Paul Johnston, Mitch Morehart, and Mark Smith. 2001. *Agri-Environment Policy at the Crossroads: Guidepost on Changing Landscape.* USDA-ERS Agricultural Economic Report no. 794. Washington, DC: Economic Research Service.

Claassen, Roger, and Richard D. Horan. 2000. Agri-Environmental Payments to Farmers: Issues of Program Design. *Agricultural Outlook* (July–August): 15–18. United States Washington, DC: U.S. Department of Agriculture, Economic Research Service.

Cox, Craig. 2001. Conservation as After-thought or Centerpiece. *Conservation Voices* (December–January): 2. Ames, IA: Soil and Water Conservation Society.

Dosskey, Michael. 2001. Toward Quantifying Water Pollution Abatement in Response to Installing Buffers on Cropland. *Agri-Environmental Management* 28 (5): 577–598.

Ervin, David E. 1999. Toward GATT-Proofing Agri-Environmental Programmes for Agriculture. *Journal of World Trade* 33 (2): 63–82.

Esseks, J. Dixon, and Steven E. Kraft. 2002. What Types of Government Assistance Are Important to Owners of Urban-Edge Agricultural Lands? Findings from Surveys Conducted from August 2001 to February 2002 in California, Michigan, New York, Texas, and Wisconsin. DeKalb, IL: American Farmland Trust.

Goolsby, Donald A., William A. Battaglin, Gregory B. Lawrence, Richard S. Artz, Brent T. Aulenback, Richard P. Hooper, Dennis R. Keeney, and Gary J. Stensland. 1999. *Flux and Sources of Nutrients in the Mississippi-Atchafalaya River Basin.* Topic Three Report submitted to the White House Office of Science and Technology Policy, Committee on Environment and Natural Resources, Hypoxia Working Group, May.

Heimlich, Ralph. 1994. Targeting Green Support Payments: The Geographic Interface between Agriculture and the Environment. In *Designing Green Support Programs; Policy Studies Report,* no. 4 (December): 11–54. Greenbelt, MD: Henry A. Wallace Institute for Alternative Agriculture.

Heimlich, Ralph. 2003. *Agricultural Resources and Agri-Environmental Indicators, 2003.* Agricultural Handbook AH 722, December. Washington, DC: U.S. Government Printing Office.

Hellerstein, Daniel, Cynthia Nickerson, Joseph Cooper, Peter Feather, Dwight Gadsby, Daniel Mullarkey, Abebayehu Tegene, and Charles Barnard. 2002. *Farmland Protection: The Role of Public Preferences for Rural Amenities.* Washington, DC: USDA Economic Research Service Agricultural Economic Report no. 815.

Helms, J. Douglas. 2003. Performance-Based Conservation: The Journey toward Green Payments. *Historical Insights,* no. 3 (September). Washington, DC: Natural Resources Conservation Service.

Hoppe, Robert, ed. 2001. *Structural and Financial Characteristics of U.S. Farms:* 2001 Family Farm Report. USDA-ERS Agriculture Information Bulletin no. 768. Washington, DC: Economic Research Service.

Horan, Richard, James Shortle, and David G. Abler. 1999. Green Payments for Nonpoint Pollution Control. *American Journal of Agricultural Economics* 81 (5): 1210–1215.

Hrubovcak, James, Utpal Vasavada, and Joseph E. Aldy. 1999. Green Technologies for a More Sustainable Agriculture. Agricultural Information Bulletin no. 752. Washington, DC: U.S. Department of Agriculture, Economic Research Service.

Hummon, Lisa, and Frank Casey. 2004. *Status and Trends in Federal Resource Conservation Incentive Programs, 1996–2001.* Working Paper no. 1. Washington, DC: Defenders of Wildlife, Conservation Economics Working Program.

King, Dennis M., and Peter J. Kuch. 2003. Will Nutrient Credit Trading Ever Work? An Assessment of Supply and Demand Problems and Institutional Obstacles. *Agri-Environmental Law Review: News and Analysis* 33 ELR 10352.

Kline, Jeffrey, and Dennis Wichelns. 1996. Measuring Public Preferences for the Environmental Amenities Provided by Farmland. *European Review of Agricultural Economics* 23:421–436.

Offutt, Susan, Betsey Kuhn, and Mitchell Morehart. 2004. Devolution of Farm Programs Could Broaden States' Role in Ag Policy. *Amber Waves* (November): 14–21. Washington, DC: U.S. Department of Agriculture, Economic Research Service.

Organization for Economic Cooperation and Development (OECD). 2001. *Multifunctionality Towards an Analytical Framework*. Paris: OECD Publications.

Reinsel, Bob. 1998. The Agricultural Resource Management Study: Serving the Information Needs of Agriculture. *Agricultural Outlook* (September): 7–9. Washington, DC: U.S. Department of Agriculture, Economic Research Service.

Ribaudo, Mark, Ralph Heimlich, Roger Claassen, and Mark Peters. 2001. Source Reductions vs. Interception Strategies for Controlling Nitrogen Loss from Cropland: The Case of Gulf Hypoxia. *Ecological Economics* 37 (May): 183–197.

Schweikhardt, David B., and William P. Browne. 2001. Politics by Other Means: The Emergence of a New Politics of Food in the United States. *Review of Agricultural Economics* 23 (2): 302–318.

Stephenson, Kurt, Leonard Shabman, and James Boyd. 2005. Taxonomy of Trading Programs: Concepts and Applications to TMDLs. In *TMDLs: Approaches and Challenges*, ed. Tamim Younos, 253–286. Tulsa, OK: PennWell Books.

Soil and Water Conservation Society (SWCS). 2000. Reforming the Farm Bill: Ideas from the Grassroots. Ankeny, IA: Seeking Common Ground for Conservation: An Agricultural Conservation Policy Project.

U.S. Department of Agriculture (USDA). 1981. *A Time to Choose: A Summary Report on the Structure of Agriculture*. Washington, DC: USDA.

U.S. Department of Agriculture–Economic Research Service (USDA-ERS). 1997. *Agricultural Resources and Agri-Environmental Indicators, 1996–97*. Agricultural Handbook no. 712, December. Washington, DC: U.S. Government Printing Office.

U.S. Department of Agriculture–Economic Research Service (USDA-ERS). 2000. *Farm Resource Regions*. Agricultural Information Bulletin no. 760. Washington, DC: Economic Research Service.

The Prospects and Limits of Antitrust and Competitive-Market Strategies

Peter Carstensen

Most farmers and ranchers lack power as either buyers or sellers. They are not able to resist price increases from suppliers or price reductions imposed by their large customers. In addition, the increasing use of contractual and other devices restrict the freedom of action of the farmer in selecting the mix of products and services used in production as well as deny them equal and open access to markets for the sale of products (O'Brien 2004). Competitive market issues involve two related categories of competition policy: traditional antitrust law and market-specific regulation. Each category presents significant issues in the context of the rapidly changing technology, market structure, and business practices in agriculture. Moreover, applying antitrust to such markets requires a focus on an issue that has not had sustained attention in the past: buyer power.

This chapter has several goals. First, it defines and differentiates between antitrust and market-facilitating regulation. Second, it briefly describes and illustrates the competitive issues facing U.S. agriculture today. Third, it provides a critique of existing patterns of antitrust enforcement, including the underappreciated issues raised by buyer power, and the limits of antitrust as a force in reforming markets, especially those in which agricultural commodities are sold. Fourth, it critically examines the current market-facilitating regulations applicable to agriculture.

Antitrust Law and Market-Facilitating Regulation

Public policy in the United States employs two strategies to accomplish the goals of fair, efficient, and accessible competitive markets: antitrust and market regulation. Both are important elements of an overall competition policy, but each serves an overlapping yet distinct set of goals.

Antitrust Law

The Sherman and Clayton acts seek to preserve competitive markets. The Clayton Act's section 7 prohibits mergers that may substantially lessen competition (Ross 1993). The primary focus of merger law today is on combinations that eliminate actual competition and reduce the number of major competitors in a market to two or three. On the other hand, lesser reductions in competition as well as mergers affecting potential competition or those likely to cause other indirect injuries are rarely challenged. Moreover, if the merging firms compete in multiple markets, the emphasis in settlements is on partial divestiture that only eliminates the effects of the merger in the specific market, but allows the combination overall, even if that eliminates some competition in a number of other markets (ibid.).

The Sherman Act interdicts collective action by firms (conspiracy, contract, or a combination) that both restrains competitive freedom and lacks legitimate economic justification (section 1), and exclusionary and abusively exploitative conduct by monopolists (section 2) (ibid.). Restraint of trade analysis employs two standards. Naked restraints of competition such as pure price fixing or market allocation are illegal per se. Second, restraints that are or appear to be functionally related to the facilitation of a legitimate transaction or joint venture among the parties are evaluated under the "rule of reason" that balances the gains and justifications for the restraint against any competitive harm it may create (ibid.). While the enforcement effort against cartels remains strong, the current enforcement against other unnecessary restraints has weakened as a result of a lack of effort by the enforcers and judicial hostility, although a few decisions show continued vitality in the area (e.g., *U.S. v. Visa*).

Section 2's prohibitions apply only if a firm has a monopoly ("monopolization") or a significant probability of obtaining one ("attempt to monopolize"). In such cases, exclusionary practices that lack a legitimate business justification are forbidden. In addition, exploiting customers or suppliers through the misuse of market power (e.g., tying the sale of a monopoly product to a requirement to buy some other product) can be illegal, although the scope of this prohibition is unsettled. The courts have recently shown more sympathy for cases challenging such conduct (e.g., *U.S. v. Microsoft*; *Conwood v. U.S. Tobacco*; *LaPage's v. 3M*), but the U.S. Supreme Court, at the behest of the U.S. Department of Justice, responded with a restrictive restatement of what constitutes unlawful exclusion (*Verizon v. Law Offices of Curtis V. Trinko*).

The antitrust laws assume that markets in general are workably competitive and produce socially desirable results. Hence, antitrust law polices the margins of conduct and structure. Antitrust law cannot rewrite the rules or revise the general institutions within which such markets operate, because the courts lack the authority under antitrust law and do not have the administrative capacity to create or enforce general regulations (Sullivan and Grimes 2000, 10–11).

Market Regulation

Markets involving any level of complexity in transactions must be constituted through some set of rules, including ones defining when and how the participants will transact with each other. The goals for market-constituting regulation, whether public or private in origin, should be equitable access, fair terms, nondiscrimination, and good information for all participants.

When such regulations exist, all participants have a reasonable chance to participate in the market. But any market-constituting regulation will inevitably allocate economic opportunities. Hence, such regulation can be hotly contested, and the process subject to substantial pressure from interested parties and groups. Actual market constitutions may therefore vary substantially from the ideal, especially if the allocation of power among participants is unequal either because some are much larger than others or because some group is better organized. Particularly in response to such situations, Congress and the states have created a number of specific statutory systems to protect the less powerful parties in their relationships with powerful customers or suppliers (e.g., Automobile Dealers Day in Court Act, Petroleum Marketing Practices Act, Securities and Exchange Act, and the Wisconsin Fair Dealership Law). A central goal of such regulations is the need to ensure the equitable treatment of all participants in an economic process (Macaulay 1966). Because such legislation is also vulnerable to special interests, it can result in "rent-seeking" or "protectionist" regulation. Critics focus on the mis-allocation of resources that may result from such regulation (Wiley 1986). Such criticism, however, ignores the underlying need for the positive regulation of specific practices in many market contexts.

Agricultural product markets have long histories of abuses, informational disparity, and strategic advantage. Hence, the public regulation of such markets dates to the earliest days of the English common law (Letwin 1965, 18–52).

Starting at the beginning of the last century in this country, Congress adopted a series of statutes intended to provide a legal framework for agricultural product markets (Commodities Exchange Act and Grain Inspection Act). It federalized grain inspection, grading, and weighing to reduce the power of the commodity exchanges in overseeing grain marketing. This made it possible for buyers to deal directly with the local elevator, bringing effective competition closer to the farm gate.

A similar pattern occurred in livestock markets. First, there was the government grading and inspection of meat that reduced some of the barriers to entry into meat processing. Then, Congress adopted the Packers and Stockyards Act (PSA) that provided for the direct regulatory oversight of and prohibited unfair discriminatory practices by stockyards and slaughterhouses. Section 202(a) of the PSA forbids packers from using "any unfair, unjustly discriminatory, or deceptive practice or device" without requiring that such practice or device also have an anti-competitive effect or intention. Similarly, section 202(e) condemns "any course of business" that has "the purpose" or "effect of manipulating or controlling prices" as well as such acts when they cause monopoly or restrain competition in the market. The PSA expressly authorized the adoption of regulations to ensure a fair, open, and accessible market for livestock. The PSA has a goal: it instructs the secretary of agriculture to regulate to protect producers as well as buyers from unfair and discriminatory conduct.

In dairy, milk marketing orders authorized by the Agricultural Marketing Agreement Act created a different approach to the problem of equitable market access. Each order area offers a different premium for fluid milk and standard base prices for Grade A milk used for other purposes. These prices are then averaged based on the relative use of milk in the order area to create a "blended" price that is the basic payment to all

Grade A dairy farms that participate in the order. In theory, all partici-pating farmers in the order area are to receive the same base price re-gardless of the use made of the milk from specific farms. The order system thus theoretically provides a method of capturing the higher value arising from some uses of milk and sharing the resulting revenues. In practice, the order system has resulted in differential prices unrelated to costs or efficiency of production that favor some regions of the country over others because of economically unjustified price differentials (Crop and Jesse 2004). More troubling, the order system limits which Grade A farmers can participate in the blended price system, and authorizes large cooperatives to move milk between order areas and reallocate the result-ing revenue as they see fit (7 USC 608C[5][F]). These practices facilitate market manipulation by allowing large cooperatives to pay excessive prices in competitive regions. This distorts the market process and results in unjustified discrimination among milk producers.

The Agricultural Fair Practices Act forbids refusals to deal based on the potential seller's participation or refusal to participate in an "agricul-tural organization" such as a cooperative. Congress found that buyers were refusing to deal with producers who organized cooperatives to bar-gain on their behalf, and on the other hand, claimed that some coopera-tives had compelled their customers to refuse to deal with nonmembers (7 USC 2301). Yet the statute has been largely ineffective because it excuses any refusal where the buyer has a plausible pretext for its action (7 USC 2304).

The Capper-Volstead Act exempts agreements among agricultural cooperatives as well as between cooperatives and their members from antitrust law. Some cooperatives, especially in dairy, are basically bar-gaining organizations that represent their members in dealings with buyers. Such collective action is intended to serve as a counterweight to the power of large buyers. Absent an exemption, such collective action would be a violation of antitrust law as a conspiracy to eliminate com-petition. Unfortunately, the authority of the secretary of agriculture to regulate cooperatives does not provide for oversight of the conduct of these organizations, even though the size and secrecy of the largest of these organizations raises some serious concerns.

The common policy goals underlying these statutes are fair, accessible, and efficient markets. Nevertheless, specific legislation responds only to particular issues and problems. There is no systematic statutory frame-work to constitute workable agricultural product markets. In contrast, in securities and commodities futures markets that face many similar risks, the federal government has established comprehensive market-facilitating statutes and regulations.

Concerns for competition and fair market processes tend to yield sim-ilar requirements. In the case of concentrated markets, however, there is some tension. To induce competitive efforts among existing dominant firms, concealing information and creating opaque market situations may induce such firms to behave in a more competitive way. Conversely, creating greater price transparency is likely to facilitate tacit coordina-tion among dominate firms provided substantial barriers to entry re-main. Still, reducing the capacity of dominant firms to engage in opportunistic, strategic behavior with respect to key inputs through reg-ulating the manner in which the market for inputs operates, offers the kind of assurance that new entrants or marginal firms seeking to expand need to encourage more active competition on the merits. These tensions underscore the complexity of the choices that must be made.

Concentration and Competition in Agricultural Markets

The two hallmarks of contemporary markets from which farmers and ranchers buy, and into which they sell, their produce are: high and increasing concentration; and new forms of strategic conduct that further frustrate competition. These phenomena are not unrelated. As markets get more concentrated, the incentives for dominant firms to engage in strategic conduct increase. The prospect of gains as well as a desire to avoid foreseeable risks posed by rivals both motivate such conduct. The following survey demonstrates that both concentration and strategic con-duct have had an increasing impact in agricultural markets.

Concentration in Agricultural Markets

The dairy industry in its various forms is a good starting point. There has been a substantial transformation of the business of milk processing. As a result of recent mergers and affiliations, Dean now processes over 30 percent of all fluid milk in the United States. The next two largest processors, Hood and National Dairy Holdings, have entered into a strategic alliance involving cross stock ownership and management. In addition, the Dairy Farmers of America, the largest dairy cooperative in the country, has contractual and/or ownership links to all three of these firms. Other elements of the dairy business are similarly or more concentrated. Kraft dominates the cheese market, buying about 30 percent of all cheese sold in the United States, and Land O'Lakes, the second largest dairy cooperative in the country, now has a similar dominance in butter production.

The livestock slaughter industry is even more concentrated. In 2003, the four largest firms in steer and heifer slaughter did more than 80 percent of that business nationally (Harkin 2004). One firm alone, IBP, did more than 32 percent. The concentration in pork that year was lower, with the four largest firms accounting for 64 percent of that business nationally (ibid.). These data understate concentration from the perspective of livestock raisers. Rarely do more than two or three of these firms compete as buyers in any region. In many areas only one major firm is effectively buying. Moreover, there is consolidation among leading firms dealing with different types of livestock. The largest poultry processor (Tyson) acquired the largest beef and second largest pork processor (IBP). Meanwhile, the largest pork processor (Smithfield), the only leading pork processor not previously engaged in beef processing, has acquired two beef processors and is now the fifth largest beef processor in the country.

In poultry, a handful of large firms dominate the field. The largest firm processes about one-third of all chickens. The markets for grain are increasingly concentrated on the buying side as a result of mergers—four firms control 60 percent of all terminal elevators and 98 percent of the Gulf Coast storage terminals (ibid.). Similarly, the largest grocers have an increasing share of all sales—the top five chains control 46 percent of all sales in the country (ibid.). This is a direct consequence of the merger mania that has afflicted this industry. In response to this, food processors have also engaged in a number of mergers. Such mergers can have a specific impact on the demand for particular farm products. For example, Nestlé's acquisition of Ralston Purina's pet food operation combined the two largest buyers of beef and poultry meal products, creating near monopsonistic buying power for these agricultural by-products (Elhauge 2001).

Concentration as a result of mergers has also increased in a number of important areas of agricultural supply, including farm equipment manufacturers (Lawton 2003). Perhaps the most troubling area on the supply-side is the rapidly concentrating biotechnology field that is central to new generations of seeds, herbicides, and pesticides. In the few years preceding 2000, one count found that the major biotech firms had engaged in sixty-eight mergers (MacDonald 2000). The pace of combinations has continued, resulting in rapidly increasing concentrations in these biotech fields (Harkin 2004). For example, Monsanto controls the most popular herbicide-tolerant soybean and has acquired control over the best sources of seed germ necessary to developing improved soybeans. Thus, it now controls its competitors access to an essential element for developing new types of seed (McEowen 2004).

Strategic Conduct in Concentrated Markets

Concentration creates incentives to exercise the resulting market power. The most direct impact is in the markets for agricultural products. With fewer and fewer buyers, producers have no choice but to accept the offers. In the cheese business, Kraft and others manipulated their purchases of cheese on the old Green Bay cheese exchange to drive down the price (Mueller et al. 1996). Recent data on the margins in meatpacking show that the margins retained by the slaughterhouses and retailers have increased substantially (Brasher 2005). This suggests that the slaughterhouses are more vigorously exploiting their buying power to depress farm prices relative to the prices they are getting from the grocery stores. Again, buying power gets reflected back toward the level of the market least able to resist.

This same buying power allows the large grocery chains to drive down the price of other products they buy without necessarily reducing the price to the consumer. The chains' insistence on slotting allowances—up-front payments before stocking a line of goods—is another example of using buying power to distort competition. As a result, greater burdens are imposed on small processors with limited product lines because they have lower volumes over which to spread these costs. Additionally, such conduct creates barriers to entry, and reduces the potential for price and product competition, further limiting the number of buyers for agricultural products (Jennings et al. 2001; Carameli 2004).

Food processors faced with price cuts or the costs of slotting fees have the market power to pass these costs back to the farmer and rancher in the form of lower prices, further reducing farm income. The farmers and ranchers are so atomistic in structure that they can neither resist nor pass on the reduction in prices. Size confers bargaining power even though it does not confer any meaningful productive efficiency (Jennings et al. 2001). Hence, when processors claim that a merger will give them greater bargaining power, they are announcing that they plan to use their newly acquired buying power to reduce the prices paid to farmers while trying to keep their prices to retailers up (Fee and Thomas 2004).

The consolidation of grain handlers has caused substantial increases in concentration at the national and regional levels. This has re-created the situation of the early 1900s for grain producers: little or no choice in the buyer. Once again, local monopsony (a single buyer able to set prices) creates a serious risk of harm to producers even if those local monopolists in turn sell into more competitive national and international markets (Carstensen 1992).

In the markets for cattle and hogs, the implication of captive supply (advanced contractual commitments to deliver animals) in livestock for efficient market operation is fourfold (McEowen 2004; O'Brien 2004). First, the favored feeder with such a contract based on a pricing grid promising prices above those prevailing in the public market has an incentive to serve its economic master because the next best option involves a substantial loss of revenue. These same buyers have an interest in manipulating the public market prices to lower both the prices of captive supplies and the price of public market purchases. A recent jury verdict found that such price manipulation cost public market sellers approximately $1.2 billion over in lost income (*Pickett v. Tyson Fresh Meats*). Yet the trial court held the conduct lawful despite the Packers and Stockyards Act's ban on price manipulation because the conduct had possible efficiency justifications, and no regulation prohibited or even regulated these practices. Second, the feeders are not well positioned to bargain effectively on the terms of the transaction. Third, buyers are under no obligation to deal with all feeders on equal terms, and can refuse to deal with any feeder for any or even no reason. High concentration on the packers' side means that such refusals will usually deny all access to the more lucrative contract market or even eliminate a disfavored operator from the business entirely. Fourth, captive supply contracts are "secret" because the packers have successfully insisted that the terms are confidential business information. Farmers thus lack the information necessary to evaluate the reasonableness of the terms that they are being offered. The capacity for the slaughterhouses to exploit feeders as well as contribute to inefficiency and market failure is a result of both the concentration of these markets and the current legal regulation, or lack thereof, of livestock-buying practices.

The most extreme example of what can happen is found in poultry. Once there was an open market in which growers could sell their chickens and turkeys. Today, there is no longer a spot or public market for general production. All production takes place under contracts that impose conditions on the growers that essentially transform them into contractors providing certain services for the integrated poultry processor (Roth 1995). In fact, an Oklahoma Attorney General Opinion (2001) claims that many contracts for the production of crops and livestock are now contracts of adhesion that may in fact reduce independent farmers to the position of employees.

With greatly increased concentration on the buying side, the risk is that the public markets, already subject to serious manipulation, will disappear entirely. Firms with large market shares have significantly greater incentives to limit the scope of the public market and seek the strategic advantages that come from contractual integration. Conversely, in a market with many buyers and sellers, the incentives to manipulate and frustrate the market process are balanced by having a reliable public market process available to all participants. The costs that result from the failure to maintain a viable public market fall on farmers and ranchers as well as consumers (Rogers and Sexton 1994; Sexton and Zhang 2001).

In biotechnology, exploitative contract terms and systems of rewarding dealers foreclose future competition while also extracting all economic gain from the farmer. For example, Monsanto also uses contracts based on its patent rights to foreclose farmers from replanting the seeds from the soybeans, cotton, and canola that they have raised. To continue using Monsanto's seeds, a farmer must buy new seed each year in addition to paying Monsanto's license fee (*Monsanto v. McFarling*). The costs of planting a soybean crop more than double as a result. Moreover,

Monsanto can use these same contract rights to control the resale of the crops produced with its patented genes.

Antitrust and Agricultural Markets

The Goals of Antitrust

The Supreme Court has said that the concern of antitrust is protecting "competition, not competitors" (*Brown Shoe v. U.S.*, 320). But the meaning of competition has changed over time. Three general themes run through the discussion of the appropriate focus of antitrust (Kaysen and Turner 1959, 11–20). One is a populist theme of anti-bigness in which the objective of the law is to challenge and control enterprises that are large while protecting those that are small, but this theme has never been dominant in antitrust law as enforced. A second and more important theme is that antitrust serves as part of a social and political agenda of retaining a democratic economic system. The fundamental concept is that dispersed economic power is essential for social mobility and political equality. This strand of antitrust law has found articulation in decisions down to recent times.

Antitrust laws in general and the Sherman Act in particular are the Magna Carta of free enterprise. They are as crucial to the preservation of economic freedom and the free enterprise system as the Bill of Rights is to the protection of fundamental personal freedoms (*U.S. v. Topco Associates*, 610).

The political support for dispersed economic power as a social goal declined during the late 1970s and greatly atrophied thereafter as a result of both the fall of the Soviet system and the rise of a globalized economy in which corporate size was equated with national economic survival.

The dominant contemporary focus for antitrust is on economic efficiency. The economic efficiency criterion in turn is often used as an explanation for not interfering with market structure or conduct (see, e.g., Arthur 1994). The assumption is that markets are generally efficient unless there is strong evidence that the market participants are abusing or might abuse their market power to exploit consumers or suppliers (Easterbrook 1984). There is a good deal of ambiguity as to the type of efficiency that should be of concern. In some contexts it is productive efficiency (claims that particular practices or combinations of firms will have lower production/distribution costs). In other contexts the focus is on the even more illusive concept of allocative efficiency (the overall higher value production of goods and services relative to the total social cost). Both of these concepts of efficiency are inherently static and compare two states of the world without asking about the potential for change (Carstensen 1983).

In some antitrust policy discussion, there has been a concern for dynamic efficiency: the most efficient way for economies to innovate and grow. Hence, even among those who take economic issues as the central concern of antitrust, there is much room for differing evaluations of the merits of particular policies (ibid.).

Competition policy like other long-term public policies should not make short-run economic efficiency a central criterion (see Driesen 2003). Experience teaches that there are many ways to achieve such efficiency. Instead, policymakers and competition law enforcers should seek to achieve other essential goals of public policy through organizing economic activity and market relationships. For social and political reasons as well as a long-run interest in maintaining the dynamics of the economy, large concentrations of control over specific markets or market sectors are undesirable.

Moreover, it is rare in an economy as vast as ours that high concentration is necessary to achieve desirable efficiency. Although some have argued that high concentration is necessary for significant efficiency gains, in fact it is not essential for efficiency in processing agricultural products. For example, the optimal plants for hog or beef processing require only 2 to 4 percent of the total national volume. Even if there were some further efficiency from multiplant operation, the market could easily sustain seven to ten efficient processors with two or three plants in both pork and beef. Such a structure would create a much more competitive context for ranchers (Carstensen 2000).

This is not to suggest ignoring the questions of economic efficiency and minimizing the costs of production. Those are always threshold considerations. Rather, the suggestion, supported by many decades of experience, is that the market process can usually find ways to achieve real efficiency without having to sacrifice other important goals. The inherent efficiency-seeking character of workably competitive markets and the innovative capacity of market actors combine to achieve efficient results under a wide variety of legal regimes.

Applying Antitrust to Agriculture

The Supply Side On the side supplying agriculture, the relevant analysis is the same as that for any other consumer market. Farmers and ranchers are no different from other types of buyers. The issues that exist are those of the general definition and application of standards for evaluating cases.

The government in its enforcement decisions over the last twenty-five years and the courts in theirs have been too lax in allowing increased concentration, and too willing to accept questionable conduct. There has been a marked increase in tolerance for mergers that create significant concentration on the selling side despite substantial evidence that such concentration has no positive effect on efficiency. Also, there has been a willingness to accept claims that restrictive agreements might nevertheless serve some efficiency-enhancing function in the market. This really harms competition, and more critically, undermines the capacity of the market to provide consumers with a wide range of goods and services. A greater skepticism about the putative benefits of size and integration would be a healthy reaction to the recent tolerance of both mergers and restraints of trade.

Regrettably, there has been little scholarship that has followed up on and investigated the actual consequences of structural changes in markets as well as situations in which antitrust law blocked change. The few studies of concentration (see, e.g., Weiss 1989) suggest that increased concentration generally results in higher prices and no efficiency gains. Also, the failure to block the massive consolidation of the seed industry in the 1990s likely has had adverse competitive effects with respect to entry, innovation, and the preservation of competing seed lines, and resulted in significant increases in prices to buyers. A study on the impact of monopoly power in cotton and soybean seeds found that the patent holder was able to appropriate most of the economic surplus created by the new products (Falek-Zepeda et al. 2000a, 2000b). This greatly alters the historic pattern in which farmers earned substantial premiums by being innovative and adopting new technology. The historical records, furthermore, show that antitrust interventions do not cause harm to the economy even if the particular decision had a questionable economic rational (Carstensen 1989). On the other hand, with some notable exceptions including tin cans and copiers, it is difficult to identify specific cases where antitrust litigation alone made a major difference (Mckie 1955; Tom 2001). More studies on the impacts of changes in the structure and conduct of supply markets would help inform policy choices. But well-counseled business decision makers consider antitrust law in making decisions about mergers and specific conduct. Hence, observed changes in market structure and conduct are usually the result of the interaction of complex economic and legal forces. Technological change and changes in other legal regulations also occur along with antitrust intervention further complicating the analysis.

The Selling Side   When farmers and ranchers sell goods into concentrated markets, the analysis is of buyer power. There is a long-standing recognition in case law and economics that there are clear risks of anticompetitive consequences from increased buyer power as from increased seller power (Blair and Harrison 1993). In 2000 and 2001, three significant court of appeals decisions reemphasized the dangers of buyer power to the overall competitive operation of the market. (*Toys R Us v. FTC*; *Knevelbaard Dairies v. Kraft*; *Todd v. Exxon*). Congressional hearings on the effect of slotting allowances have highlighted comparable harms to competition in the food distribution system (see Jennings et al. 2001). Indeed, tobacco farmers collected substantial damages in settlement of their claims that the major cigarette companies had engaged in a conspiracy to depress the price of tobacco (*DeLoach v. Phillip Morris*), and blueberry producers in Maine got a substantial judgment against the processors who had conspired to depress the prices paid for that crop (*Pease v. Jaspar Wyman & Son*).

Despite evidence of the significance of buying power, its implications have not been a primary focus of antitrust analysis. Some work in agricultural economics has addressed the overall implications in theory and empirically, but it has not been linked directly to changes in market structure or the impact of particular practices by buyers (Rogers and Sexton 1994; Sexton and Zhang 2001). Nevertheless, experience to date strongly suggests that different standards might be important in order to evaluate correctly the competitive implications of buying power (Carstensen 2004). In at least some circumstances, firms with relatively low market shares have demonstrated significant buying power (Fee and Thomas 2004). Moreover, the beef-processing mergers of the 1980s show that even where conventional, selling-side analysis would indicate no serious competitive problems (e.g., *Cargill v. Monfort*), there have been serious adverse competitive effects on the buying side.

In developing selling-side standards, several considerations appear important. First, a producer may need access to a large number of downstream buyers. When upstream producers need a key retail outlet or many retail outlets to achieve the necessary volume and efficiency, the failure to get access results in significant costs. In such situations, every buyer with a significant share—10 percent or more of the retail market —has substantial power over the supplier (e.g., *Toys R Us v. FTC*). This suggests that merger analysis ought to be attentive to the creation or expansion of such power even where the market is not concentrated by conventional standards. Similarly, restrictive terms or special obligations on sellers in such contexts should be subject to stricter scrutiny under restraint of trade analysis even where the market shares of buyers imposing these restraints seem low.

Second, when a buyer imposes lower prices on its supplier, that supplier is likely to seek to reduce its input costs. For example, when grocery stories reduce what they pay for goods, the processors have every incentive to pass those lower prices up the supply line (Fee and Thomas 2004). A downstream price cut on inputs will have a ripple effect and ultimately impact upstream parties without the power to respond. For instance, major cheese buyers drove down the price of cheese, and cheese makers in turn reduced the price of milk to the farmer (Mueller et al. 1996). The implication for both merger and restraint analysis is that adverse competitive effects may occur in the second or third tier of supply markets. Therefore, transactions require more comprehensive investigation, and the policy goal of antitrust—to focus on protecting the competitive process—should be emphasized rather than whether immediate competitors are harmed.

Third, where the producer has relatively few choices of buyers even though further downstream the ultimate consumer may have a number of choices, the buyer has power in the upstream market to discriminate among and impose onerous burdens on its suppliers. Some agricultural product markets fit this model. The local grain elevator or slaughter-house, for example, will have substantial advantages over its supplier be-cause the supplier has few, if any, options. Mergers or restraints that increase the capacity for a firm to engage in such conduct are anticompe-titive. Similarly, combinations that increase the capacity of a firm to exploit buying power or agreements selectively raise important anticom-petitive concerns.

There is no clear empirical or theoretical map of the contexts in which such dangers are enhanced, or conversely, where countervailing market facts would make such harms unlikely. The Federal Trade Commission's (1994) "Merger Guidelines" provide a detailed analysis of the circum-stances, on the selling side, under which an inference of adverse compet-itive effect is or is not plausible, but offer no comparable guidance on the buying side. The same is true of the Federal Trade Commission's (2000) "Guidelines on Collaboration among Competitors."

Effective antitrust enforcement requires a greater appreciation and deeper analysis of buyer power. This would better inform all antitrust enforcement, but would be particularly important for cases involving the sale of agricultural products. This, then, is an argument for expand-ing the scope of carefully developed, general antitrust doctrine to take better and more consistent account of the issues involved in examining the buying side of cases.

With regard to the structure and conduct of agriculture product mar-kets, the suggestion is that enforcement was particularly lax because nei-ther the agencies nor the opponents of those combinations and conduct had a strong theoretical and empirical basis for the intuition that, for ex-ample, the creation of the new Dean milk and its relationship through the Dairy Farmers of America with Hood and National Dairy Holdings would cause serious supply-side problems. Similarly, many current buy-ing practices in livestock markets appear to be anticompetitive and ex-ploitative. New entry is difficult and the competitiveness of upstream supply markets are undermined without any economic efficiency justifi-cation. The problem from the standpoint of current antitrust thinking, again, is that the theory and empirical evidence of such harms have not yet emerged from academic study.

The Limits of Antitrust in Market Facilitation

Antitrust is focused on the overall competitiveness of markets. As such, it has little interest in conduct that reduces the relative income of some set of market participants unless that is seen as having a direct effect on the competitiveness of the markets involved. Antitrust law, therefore, has not played a prominent role in addressing the kinds of direct monopsonistic conduct problems that have emerged in livestock and poultry markets (Taylor 2003).

Pure wealth transfers are not identified as harming the workable competition in those markets. Specifically, the current contracting system for raising poultry often results in near serfdom for the farmer and transfers disproportionate risks. Farmers who have disputes with the integrator also are usually constrained by unfair arbitration agreements that seriously limit their ability to obtain just resolutions (Roth 1995). All of this in effect transfers the farmer's wealth to the integrator (see Taylor 2003), but conventional antitrust is unlikely to regard such conduct as anticompetitive in itself. A broader antitrust perspective should look at the ways in which processors locate their facilities to minimize overlap, and maximize the opportunity to exploit farmers and find a conspiracy to allocate input markets. Even so, the problem for antitrust enforcement would be a remedy: Can a court effectively regulate the location of poultry-processing facilities? Such regulation is not a traditional part of antitrust because it would in fact usurp the market function of individual enterprise decisions.

Market-Facilitating Regulation in Agriculture

The present legal framework for agricultural product markets has three major deficiencies: the statutory authority is a patchwork, the USDA has failed to use its powers within the areas over which it has authority to develop appropriate market-facilitating regulations, and the USDA's enforcement of even the existing rules has been ineffectual.

The Statutory Patchwork

The authority of the secretary of agriculture to police the fairness and equity of treatment in agricultural markets is limited and different from product market to product market. For example, the PSA applies only to meatpacking. No comparable authority exists to police grain markets at the farm level. There is authority over dairy contracts whenever a federal marketing order applies, but such orders do not apply in some key markets such as California, which has its own milk order system. Moreover, the AFPA does not confer any authority on the secretary of agriculture to issues rules that might impose requirements on the market for fair conduct. As market structure and conduct akin to that in livestock and poultry markets come to dominate other sectors, it will be increasingly important that there be regulations to ensure fairness in pricing and equal access to market opportunities for all farmers.

While the broad public policy of Congress—consistent across a wide range of specific pieces of legislation—is to facilitate an open, competitive market in agricultural products, the implementation of that policy is deeply flawed. Real reform should rest on the underlying basic structure that Congress has adopted: facilitate efficient transactional markets; limit the scope of strategic behavior and incentives to engage in discriminatory conduct; require open access to all major methods of selling farm products, whether transactional or contractual; and mandate the full and timely disclosure of relevant information to all market participants.

The first element is the most crucial one. The USDA needs to be much more proactive in developing new grading standards and certification systems so the transactional market can provide a place in which buyers could readily find the kind as well as quality of farm product that they seek (see Koontz 2000). It is not enough to ban bad practices; the government must take the initiative to modernize the spot market and related market transactions to facilitate desired transactions because public markets in agricultural products will not happen on their own in equitable and fair ways. The role of government is to restore the balance and facilitate the equitable development of the market.

The law should also minimize or eliminate the unfair, strategic, inefficient conduct of powerful buyers. In a lawless market, economic power is unchecked. That which is rational for individual, powerful economic actors is not necessarily equitable to the parties on the other side of the transaction or in the best interest of economic efficiency in the long run.

One of the most important ways to resolve the problems of exclusion and discrimination is to constitute markets in ways that allow all producers to have access to all significant methods of selling products.

Thus, if captive supply in beef or pork is necessary, all feeders should have the right to tender animals in response to public bids by the processors. Such open access was a central feature of the earlier regulation of grain and livestock markets that has been lost in more modern times.

Lastly, it is a basic economic insight that information is power. Currently too much price making and too many transactions go undisclosed. This leads to market inefficiency, and facilitates discrimination against and the exploitation of sellers who lack good information. As Congress has come to rely more on the marketplace to set prices and allocate demand for agricultural supplies, there is an increased need to review this legislative thicket and identify a systematic set of rules to govern the market process in all agriculture products.

Not Adopting Regulation Authorized by Law

The USDA has failed to use the authority given to it under the PSA, Capper-Volstead, the marketing order law, and other legislation to frame rules that in fact facilitate open, fair, and accessible markets where it could do so. This is a serious weakness. Congress cannot write and revise regulations that are fine-tuned to the needs of specific markets. The Securities and Exchange Commission, for example, has used its mandate of regulating public capital markets to develop an effective, comprehensive, and responsive framework within which such markets operate. The USDA, at the other extreme, has been unable even to propose relevant regulations, let alone adopt them. This is true of the PSA. In September 2000, the USDA convened a meeting of experts on captive supply issues in livestock markets. The experts did not agree on much, but they were in agreement that packers should never be permitted to use as the basis for pricing their own price for cattle on the day of delivery. Such a method of pricing was unduly vulnerable to manipulation and was unnecessary for any legitimate pricing need. Yet the USDA neither proposed nor adopted even such a basic prophylactic rule. In relation to milk, the law requires all marketing orders to have rules barring unfair competitive acts and practices (7 USC, sec. 608c[7][A]). Yet so far as anyone can tell, this authority has never been used, despite real concerns about the conduct of large dairy cooperatives.

Left to their own devices, large buyers will force contracts of adhesion on to producers. (Oklahoma Attorney General Opinion 2001) Such contracts often impose unfair and inequitable arbitration terms that effectively deny the producer all recourse. Confidentially clauses keep farmers and ranchers from sharing information that would make them more sophisticated decision makers. Even price reporting is unavailable where buyers are highly concentrated. These facts mandate regulations to ensure the equitable treatment of sellers in such markets. From the perspective of preserving and improving an open, fair, efficient, and informed transactional market, it is frustrating that the USDA fails to use its powers. Such efforts would simultaneously highlight the gaps in the statutory framework for facilitating fair, open, and efficient markets for agricultural products.

The Lack of Enforcement of Existing Rules

Regulations, however good, have little effect if there is no enforcement. While farmers and ranchers can bring individual cases or class actions, such efforts are time-consuming, and will focus more on specific private concerns and less on the broad public interest in ensuring open and fair markets. Thus, effective public law enforcement is essential to the creation as well as maintenance of fair and open markets. This is the lesson of antitrust and securities laws. Yet there is widespread recognition that the USDA has failed badly in its responsibilities to police and enforce rules that do exist. These persistent failures are the object of bipartisan concern in Congress, and a source of great frustration to farmers and ranchers who look to the USDA to protect their interests.

A major underlying cause for this enforcement failure is the disjointed structure of the USDA's own enforcement efforts. While one division deals with meat and grain, another deals with cooperatives, and still others focus on market information. There is little or no coordination between the divisions on policy or enforcement. A reorganization of the administration of the USDA's responsibilities would be an important contribution to the effectiveness of its regulation of agricultural markets.

Conclusion

U.S. agriculture is at a crossroad. Increasing concentration and radical changes in buying methods have created serious problems (O'Brien

2004). These problems are likely to become worse without more vigorous antitrust enforcement and necessary market-facilitating regulations combined with the effective enforcement of those rules. A reliance on markets cannot resolve all the problems facing agriculture. But Congress has wisely elected to rely as much as possible on markets as the best mechanism to organize the movement of food and fiber from the farmer to the final consumer. Given that fundamental choice, the Congress, in writing laws, and the USDA, in enforcing them, must better perform the tasks of facilitating a competitive and efficient marketplace, while also ensuring access and fair terms to producers. At the same time, antitrust enforcers need to be much more attentive to the special issues created by buyer power. This analysis should yield a more activist antitrust enforcement in agricultural markets because those markets are particularly likely to be subject to harms arising from such power.

If there is real reform in the legal framework governing agricultural markets, and if antitrust takes buyer power seriously, great improvement in the competitiveness, efficiency, and fairness of those markets is possible. In open, fair, informed markets all participants, large and small, have the best chance to succeed on their merits.

References

Arthur, Tom. 1994. The Costly Quest for Perfect Competition: Kodak and Nonstructural Market Power. *New York University Law Review* 69:1.

Blair, Roger, and Jeffrey Harrison. 1993. *Monopsony: Antitrust Law and Economics*. Princeton, NJ: Princeton University Press.

Brasher, Philip. 2005. Tyson, Timing, and Politics. *Des Moines Register*, January 16.

Carameli, Leo. 2004. The Anti-Competitive Effects and Antitrust Implications of Category Management and Category Captains of Consumer Products. *Chicago-Kent Law Review* 79:1313.

Carstensen, Peter. 1983. Antitrust Law and the Paradigm of Industrial Organization. *University of California at Davis Law Review* 16:487–525.

Carstensen, Peter. 1989. How to Assess the Impact of Antitrust on the American Economy: Examining History or Theorizing. *Iowa Law Review* 74:1175–1217.

Carstensen, Peter. 1992. The Content of the Hollow Core of Antitrust: The *Chicago Board of Trade* Case and the Meaning of the "Rule of Reason" in Restraint of Trade Analysis. *Research in Law and Economics* 15:1–88.

Carstensen, Peter. 2000. Concentration and the Destruction of Competition in Agricultural Markets: The Case for Change in Public Policy. *Wisconsin Law Review* 2000:531–546.

Carstensen, Peter. 2004. Buyer Power and Merger Analysis: The Need for Different Metrics. Paper presented at the Department of Justice and Federal Trade Commission's Workshop on Merger Enforcement, February 17. Available at ⟨http://www.ftc.gov/bc/mergerenforce/presentations/040217carstensen.pdf⟩.

Crop, Robert, and Edward Jesse. 2004. Basic Milk Pricing Concepts for Dairy Farmers. University of Wisconsin, Cooperative Extension, A3379.

Driesen, David. 2003. *The Economic Dynamics of Environmental Law*. Cambridge, MA: MIT Press.

Easterbrook, Frank. 1984. The Limits of Antitrust. *Texas Law Review* 63:1–40.

Elhauge, Einer. 2001. Report to the Federal Trade Commission on Nestle-Ralston Merger for National Grange. Available at ⟨http://www.nationalgrange .org/legislation/Nestle_Ralston.htm⟩. Spring.

Falek-Zepeda, Jose B., Greg Taxler, and Robert Nelson. 2000a. Rent Creation and Distribution from Biotechnology Innovations: The Case of Bt Cotton and Herbicide-Tolerant Soybeans in 1997. *Agribusiness* 16:21–32.

Falek-Zepeda, Jose B., Greg Taxler, and Robert Nelson. 2000b. Surplus Distribution from the Introduction of a Biotechnology Innovation. *American Journal of Agricultural Economics* 82:360–369.

Federal Trade Commission (FTC) and Department of Justice. 1992. Horizontal Merger Guidelines. Available at ⟨http://www.uskoj.gov/atr/public/guidelines/ hmg.htm⟩.

Federal Trade Commission (FTC) and Department of Justice. 2000. Antitrust Guidelines on Collaborations among Competitors. Available at ⟨http://www.ftc .gov/os/2000/04/ftcdojguidelines.pdf⟩.

Fee, C. Edward, and Shawn Thomas. 2004. Sources of Gains in Horizontal Mergers: Evidence from Customer, Supplier, and Rival Firms. *Journal of Financial Economics* 74:423.

Harkin, Tom. 2004. Economic Concentration and Structural Change in the Food and Agriculture Sector: Trends, Consequences, and Policy Options. Report prepared by the Democratic staff of the Committee on Agriculture, Nutrition, and Forestry, U.S. Senate. Available at ⟨http://www.harkin.senate.gov⟩.

Jennings, Marianne, Robert Aalberts, and Stephen Happel. 2001. The Economics, Ethics, and Legalities of Slotting Fees and Other Allowances in Retail Markets. *Journal of Law and Commerce* 21:1–46.

Kaysen, Carl, and Donald Turner. 1959. *Antitrust Policy: An Economic and Legal Analysis*. Cambridge, MA: Harvard University Press.

Koontz, Stephen. 2000. Concentration, Competition, and Industry Structure in Agriculture. Testimony at Agricultural Concentration and Competition Hearing, April 27. Washington, DC: Committee on Agriculture, Nutrition, and Forestry, U.S. Senate.

Lawton, Kurt. 2003. Turning the Corner. *Farm Industry News*, February 1. Available at ⟨http://farmindustrynews.com/mag/farming_turning_corner/⟩.

Letwin, William. 1965. *Law and Economic Policy in America*. Chicago: University of Chicago Press.

Macaulay, Stewart. 1966. *Law and the Balance of Power: The Automobile Manufacturers and Their Dealers*. New York: Russell Sage Foundation.

MacDonald, James. 2000. Concentration in Agribusiness. Paper presented at the Agricultural Outlook Forum 2000, Washington, DC, February 24.

McEowen, Roger. 2004. Legal Issues Related to the Use and Ownership of Genetically Modified Organisms. *Washburn Law Journal* 43:611–659.

Mckie, James. 1955. The Decline of Monopoly in the Metal Container Industry. *American Economic Review* 45:499–508.

Mueller, Willard, Bruce Marion, Maqbool Sial, and F. Geithman. 1996. Cheese Pricing: A Study of the National Cheese Exchange. Report to the Wisconsin Department of Agriculture, Trade, and Consumer Protection Investigation into Cheese Prices. March.

O'Brien, Doug. 2004. Policy Approaches to Address Problems Associated with Consolidation and Vertical Integration in Agriculture. *Drake Journal of Agricultural Law* 9:33–52.

Oklahoma Attorney General. Oklahoma Attorney General Opinion. 2001. No. 01-17, April 11.

Rogers, Richard, and Richard Sexton. 1994. Assessing the Importance of Oligopsony Power in Agricultural Markets. *American Journal of Agricultural Economics* 76:1143–1150.

Ross, Stephen. 1993. *Principles of Antitrust Law*. Minneola, NY: Foundation Press.

Roth, Randi. 1995. Redressing Unfairness in the New Agricultural Labor Arrangements: An Overview of Litigation Seeking Remedies for Contract Poultry Growers. *University of Memphis Law Review* 25:1207–1232.

Sexton, Richard, and Mingxia Zhang. 2001. An Assessment of the Impact of Food Industry Market Power on U.S. Consumers. *Agribusiness* 17:59–79.

Sullivan, Lawrence, and Warren Grimes. 2000. *The Law of Antitrust: An Integrated Handbook*. Saint Paul, MN: West Group.

Taylor, Robert. 2003. Monopsony and the All-or-Nothing Supply Curve: Putting the Squeeze on Suppliers. Working Paper ES.6.2003, Auburn University, College of Agriculture.

Tom, Willard. 2001. The 1975 Xerox Consent Decree: Ancient Artifacts and Current Tensions. *Antitrust Law Journal* 68:967–990.

Weiss, Leonard, ed. 1989. *Concentration and Price*. Cambridge, MA: MIT Press.

Wiley, John. 1986. A Capture Theory of Antitrust Federalism. *Harvard Law Review* 99:713–789.

Statutes

U.S. Code

Agricultural Fair Practices Act, 7 U.S.C., §§ 2301 et seq.

Agricultural Marketing Agreement Act of 1937, as amended, 7 U.S.C., §§ 6086 et seq.

Automobile Dealers Day in Court Act, 15 U.S.C., §§ 1221 et seq.

Capper-Volstead Act, 7 U.S.C., §§ 291, 292.

Clayton Act, as amended, 15 U.S.C., §§ 8 et seq.

Packers and Stockyards Act, as amended, 7 U.S.C., §§ 181 et seq.

Petroleum Marketing Practices Act, as amended, 15 U.S.C., §§ 2801 et seq.

Securities and Exchange Act of 1934, as amended, 15 U.S.C., §§ 78a et seq.

Sherman Act, as amended, 15 U.S.C., §§ 1 et seq.

State Statutes

Wisconsin Fair Dealership Law, Wisc. Stat., chap. 135.

Court Decisions

U.S. Supreme Court Decisions

Brown Shoe v. U.S., 370 U.S. 294 (1963).

Cargill v. Monfort, 479 U.S. 104 (1986).

U.S. v. Topco Associates, 405 U.S. 596 (1972).

Verizon v. Law Offices of Curits V. Trinko, 540 U.S. 398 (2004).

U.S. Court of Appeals Decisions

Conwood v. U.S. Tobacco, 290 F.3d 768 (6th Cir. 2002).

Knevelbaard Dairies v. Kraft, 232 F.3d 979 (9th Cir. 2000).

LaPage's v. 3M, 324 F.3d 141 (3d Cir., en banc, 2003) *cert. denied*, 542 US 953 124 S. Ct. 2392 (2004).

Monsanto v. McFarling, 363 F.3d 1336 (Fed. Cir. 2004) *cert. denied*, sub. nom. *McFarling v. Monsanto*, 545 US 1139 (2005).

Todd v. Exxon, 275 F.3d 191 (2d Cir. 2001).

Toys R Us v. FTC, 221 F.3d 928 (7th Cir. 2000).

U.S. v. Microsoft, 253 F.3d 34 (D.C. Cir., en banc, per cur., 2001) *cert. denied*, 534 U.S. 952.

U.S. v. Visa, 344 F.3d 229 (2d Cir. 2003) *cert. denied*, 543 US 811 125 S. Ct. 45 (2004).

U.S. District Court Decisions

DeLoach v. Phillip Morris, 2001–2 Trade Cases P 73,409 (Mid. Dist. N.C. 2001).

Pickett v. Tyson Fresh Meats, 315 F. Supp. 2d 1172 (N.D. Ala. 2004), aff'd 420 F.3d 1272 (11th Cir. 2005) *cert. denied*, 547 U.S. 1040 (2006).

State Court Cases

Pease v. Jasper Wyman & Son, 845 A.2d 552 (Me. 2004).

Epilogue: Strategies for Strengthening the Agriculture of the Middle

G. W. Stevenson, Rick Welsh, and Thomas A. Lyson

The guarded optimism in this volume about renewing an agriculture of the middle in the United States is based on several emerging realities. First, U.S. consumers have growing appetites for high-quality, unique food that stands out from the undifferentiated commodities composing the bulk of our food supply. Restaurants, health care institutions, schools, corporate cafeterias, other food service enterprises, and some supermarkets are increasing their demand for food that not only has superior taste and nutritional qualities but also comes with stories identifying where and how it was grown. Many of these businesses also seek transparent supply chains built on business relationships they can trust and support. Direct-marketing farms often lack the capacity to supply the significant volumes of differentiated, high-quality food products, and farms and ranches producing commodities for global markets are not designed to provide such food differentiation. With assistance, many farmers and ranchers of the middle can shift their production from agricultural commodities to value-adding, differentiated food, and progressive leaders in sizable food companies are recognizing their interests can be met through the regeneration of an agriculture of the middle.

Second, a growing interest in domestic fair trade is developing in the United States. Modeled after the international movement to fairly trade food such as coffee, bananas, and chocolate, the domestic fair-trade movement couples environmental responsibility with fair treatment for farmers and food workers as well as equitable strategies for small business and rural community development. The pivotal roles proposed in this volume for civic agriculture, values-based food supply chains, and farmer collective bargaining are congruent with a domestic fair-trade agenda. Such economic approaches can be both good business and ethical.

Third, the inevitable rise in the cost of petroleum will reward diversified agricultural systems that work with natural nutrient and pest cycles as well as food systems that are organized through regional supply chains. Farms and ranches that employ practices such as crop rotations, managed grazing, and other agricultural approaches that take advantage of synergistic interactions among plants and animals can reduce costs and benefit the environment by reducing or eliminating the need for petroleum-based fertilizers and pesticides. Also, preliminary research indicates that regionally based food distribution systems are more energy efficient than national and global ones, and interestingly, even more energy efficient than local food systems. Midtier food value chains as described in this volume are designed to operate at regional scales.

For the time being, there are enough farmers and ranchers (and fishers) of the middle to supply the growing market demand for differentiated food products of superior value. Both the markets and the producers who have the potential to supply these markets are in place. But for producers to successfully participate, many will need to shift their production and marketing strategies. Research and technical assistance are needed to help food producers of the middle to restrategize and retool successfully. New organizational and business paradigms that enable farmers, ranchers, and other community food entrepreneurs to become full partners and beneficiaries in these business ventures will also need to be developed. Lastly, we need supportive public policies to connect retooling agricultural producers to these growing markets. The following strategies will help meet the research, business, and policy needs of a regenerating agriculture of the middle:

1. Research and technical assistance for restrategizing and retooling agricultural producers of the middle. This research and assistance should emphasize:
• *New farming and ranching systems that reduce dependence on petroleum, produce high-quality, differentiated food, and respond to changing markets.* These biologically synergistic systems will likely be based on annual and perennial polycultures involving both plants and animals, the genetics for which will need to be researched, developed, and adapted to regional differences. The food products produced in these

farming and ranching systems will be of superior quality and guided by regularly updated market information.

• *New or regenerated socioeconomic organizations through which farmers and ranchers create production and marketing networks large enough to convey market power as well as supply significant, consistent volumes of differentiated food products.* This will involve multilateral collaborations like cooperatives and limited liability companies, and/or bilateral partnerships in which aggregating firms provide the organizational infrastructure. Collaborations between regional networks can address the challenges and opportunities associated with regional and seasonal markets. This will also require developing quality-control systems that address food safety, weather, seasonality, multiple production sites, and quality-preserving storage and distribution mechanisms.

2. Development of new business models that engage farmers and ranchers, other community food entrepreneurs, and food customers as full partners and beneficiaries. These business models should include:

• *Value chain agreements in which farmers and ranchers are treated as strategic partners, not interchangeable (and exploitable) input suppliers.* Value chain partners have a vested interest in the economic performance and social well-being of other partners. Prices should be determined on cost-plus bases, and agreements should be fair, adjustable, and set in appropriate time frames. Groups of farmers and ranchers should be able to own and/or control their brand identity as far through the supply chain as they choose. This may involve cobranding with other strategic partners in the chain.

• *Greater inclusion of food customers in food system decision making.* Successful midtier food supply chains will seriously engage educated eaters who wish to be marketed "with" rather than marketed "at." This means creating mechanisms for authentic, ongoing dialogue between food customers, producers, and marketers. Electronic discussion forums engaging eaters and producers from common geographic areas could augment store-based conversations prompted by highly informative point-of-purchase information. Building customers' interest and trust will require meaningful, enforced standards along with third-party certification for food quality and business behavior across the chain. These standards will maintain the supply chain authenticity and integrity.

3. Public policy change to support a renewed agriculture of the middle. Change is required at both the domestic and international levels:

• *Domestic policy initiatives.* Examples of supportive federal policies include sincerely enforcing the Packers and Stockyards Act to reinstate more competitive agricultural markets; tailoring risk management programs to farmers and ranchers of the middle who are reworking their business plans; expanding rural business development grants for the creation of midtier food value chains; and targeting research, education, and extension service funds to support transitions from commodity to differentiated food production. State and local policy initiatives could target farmland preservation measures as well as the food-sourcing policies of public institutions.

• *International policy initiatives.* In addition to the international supply management and price support programs proposed in this volume for major food commodities, global and fair trade, World Trade Organization compliance, international certification, and regional product identity issues will need to be addressed. For instance, the World Trade Organization should support the geographic and regional branding of agricultural products as legitimate methods of promoting farm viability as well as rural development.

This is a pivotal time for the agriculture of the middle. Farmers are aging, and midsize farms and ranches across the United States are disappearing. We probably have no more than a decade to set in motion the substantial turnaround required to renew this vital agricultural sector. Nevertheless, there are significant emerging opportunities on which to build such regeneration. It is our hope that the ideas explored in this volume will help engage those opportunities and contribute to the important work of renewing the agriculture of the middle in the United States.

Appendix: The Changing Status of Farms and Ranches of the Middle

Mike Duffy

U.S. agriculture is changing dramatically. Consolidation is occurring at all levels of production, processing, distribution, and retailing. The rapid pace of consolidation makes it difficult to get an accurate picture of the current structure of agriculture. This appendix will use two primary sources of data to focus on the changes occurring in production agriculture. The first is the 2002 USDA Census of Agriculture, and the second reflects data compiled by the USDA's ERS. It is important to remember the definition of a farm used in these data sets and acknowledge the changes that occurred in compiling the 2002 Census. These changes severely restrict comparisons with earlier years, with the exception of the revised 1997 Census numbers.

A farm is defined as "any place from which $1,000 or more of agricultural products were produced and sold, or normally would have been sold, during the Census year." This definition was first used in the 1974 Census. The data presented here contain a category for farms with sales of less than $1,000. This is because farms covered in the Census only needed the ability to sell at least $1,000, even though they may not have actually sold this amount during the Census year.

The 2002 Census augmented the normal Census mailing list with an "independent comprehensive survey of sample geographic areas." The purpose was to increase the number of names on the list, especially those of small farm operators. The USDA notes that "most farms missed on the mail list are relatively small operations." The new Census mailing list was quite successful in capturing results from additional small farms. Overall, when the original 1997 numbers were readjusted for the new list, there was a 16 percent increase in the number of farms reported. The number of farms reporting sales of less than $1,000 increased by

50 percent. This represented 45 percent of the total increase in farms reported for 1997.

The change in the sampling procedure for the Census has altered the nature of the Census data. It is arguable whether or not this was a desirable change, but it is quite obvious that given the current definition of a farm, the change did capture considerably more farms. Farms with sales of less than $1,000 make up 27 percent of all U.S. farms, operate 9 percent of the land, and account for 0.4 percent of all sales. There was a 3.9 percent decrease reported in the number of all farms. But excluding the increase in the farms with sales of less than $1,000, the decrease in farm numbers was 13 percent. Irrespective of its potential shortcomings, the Census does provide the most comprehensive examination of the current state of U.S. agricultural production and changes since 1997.

Farm Size and Numbers

Currently, there are 2,128,982 farms in the United States, operating on just over 938 million acres. The number of farms in the United States decreased by 3.9 percent from 1997 to 2002, and the land in farms decreased by 1.7 percent over the same time.

The 2002 Census presented a breakdown of farms related to size in two different ways. One was by the amount of sales, which has been the traditional way to measure size. A new categorization of farms by economic class was introduced with the 2002 Census. The difference between the two measures is that government payments are excluded in the Economic Classification. Figure A.1 compares the percentage of farms using the two classification schemes.

Notice that regardless of the classification system, most farms fall into the category of less than $1,000 in sales. Slightly more than one-fourth (27 percent) of the farms recorded sales of less than $1,000, and 20 percent of the farmers are in the economic class of less than $1,000. As expected, the percentage of farms in the smaller classes changes slightly as government payments are added, but beyond the $50,000 range there is essentially no change in the percent of farms using either classification system.

The remainder of this appendix will use the sales categorization in discussions since this is the more traditional way to discuss farm structure.

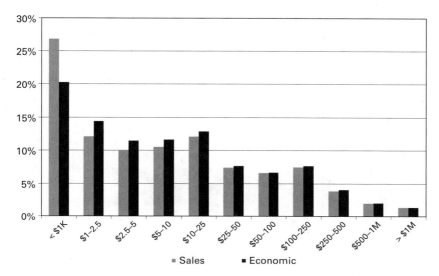

Figure A.1
Percent of U.S. Farms by Sales and Economic Class, 2002

Using economic class will increase, decrease, or leave unchanged the percent of farms relative to the sales categorization, so the impact would be ambiguous.

Figure A.2 records the changes that have occurred in farm numbers by sales class since 1997. This figure also presents the change in the percent of sales attributable to the various sales categories. Notice that the small farms (with sales of less than $1,000) increased by 37 percent, and the large farms (with sales greater than $1 million) increased by 8 percent. This illustrates the phenomenon of the disappearing middle. Figure A.2 also shows that the percentage of sales in all categories, except the largest, experienced a decrease. The percentage change in farm numbers and sales is approximately the same except for the small and large farms. For the entire production agriculture sector overall, sales were essentially unchanged, decreasing less than half of a percent.

The Census definition and listing skew the ability to make some general comparisons. This is because the average numbers include the small farms, and as depicted in figure A.2, this will have significant influence. Therefore, for the remainder of this appendix the discussion will focus on three categories of farms. The first category—sales less than $50,000—

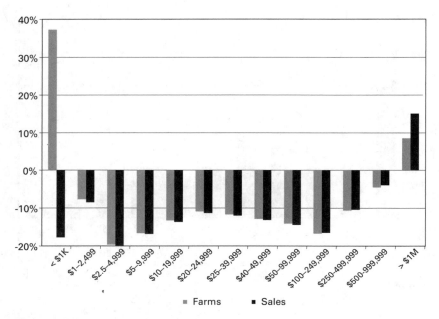

Figure A.2
Percent Change in U.S. Farms and Sales by Sales Class, 1997–2002

represents farms that are small. These farms will certainly have off-farm income. In addition, as shown in figure A.3, while this group represents almost 80 percent of the farms, it constitutes less than 10 percent of the sales. The second category of farms considered consists of those with sales between $50,000 and $500,000. Finally, the last group is made up of those farms with sales over $500,000. Any division is arbitrary and can be debated. What I am trying to accomplish with this three-way division, however, is to simplify the comparisons, and be more reflective of a noncommercial, small commercial, and large commercial triad division. The farms between $50,000 and $500,000 approximate what is being called the agriculture of the middle.

Figures A.3 and A.4 compare the relative levels of sales and the amount of land held by sales category, respectively. As expected, the farms in the smallest sales category represent the smallest percent of sales. The largest single category—those with sales greater than $500,000—had just over 60 percent of the sales. Figure A.4 shows that almost half (46 percent) of

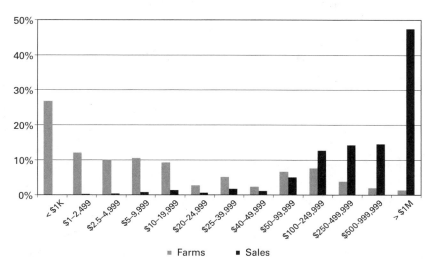

Figure A.3
Percent U.S. Farms and Sales by Sales Class, 2002

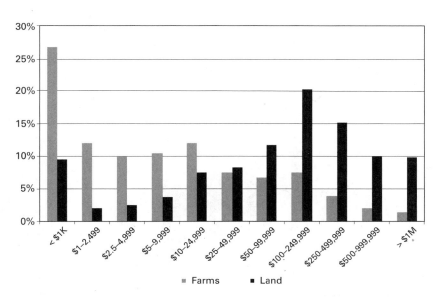

Figure A.4
Percent of U.S. Farms and Land by Sales Class, 2002

the land is held by those in the three sales categories from $50,000 to $500,000.

When the sales categories are aggregated, the 79 percent of farms with sales under $50,000 represent 6.8 percent of the total sales and 33 percent of the land. The farms with sales over $500,000 comprise 1.3 percent of the farms with 60 percent of the sales and 20 percent of the land. In addition to holding nearly a half of agricultural land, the 18 percent of enterprises designated as farms and ranches of the middle account for 32 percent of agricultural sales in 2002.

Farm Operator Characteristics

The average age of farmers was 55.3 in 2002. This was an increase of 1.3 years or 2.4 percent from 1997. Figure A.5 shows the distribution of farmers by age category. Notice that the percentage of farmers in the 35- to 44-years-of-age cohort is equal to the percentage of farmers over the age of 70. It is not clear why the percentage of farmers in the age categories between these two would be decreasing, but it is possible that this reflects the long-term effects of the 1980 farm crisis. Twenty years ago, the farmers in the lower two and most of the farms in the third age

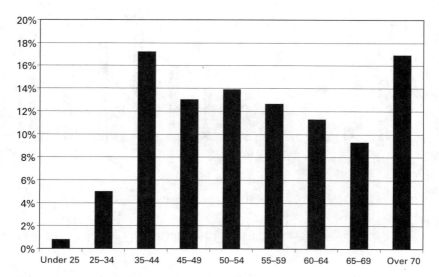

Figure A.5
Percent of U.S. Farms by Age Category, 2002

brackets would have been young children or teens. They would not own farms. The farmers in the over 70 age category would have been at the end of their farming careers, and thus better able to survey the financial turmoil that was occurring in the early 1980s. The cohort of the farmers currently in the 50- to 69-year-old brackets would have been the most likely to have been expanding up to the farm crisis. They also would have been more likely to want to expand during that time period. This is reflected in figure A.5.

The average age of farmers decreases as the size of the farm increases. Farmers in the smallest sales class were 55.9 years of age, those in the middle were 53.1, and the ones in the largest sales class were 52.3. There are many reasons for this age distribution. Many of the farmers in the smallest sales class are the retired or semiretired farmers. Additionally, farmers in the larger sales classes would more likely be at the height of their careers.

The Census customarily inquires about the principal occupation for the operator. Unfortunately, the wording of the question was changed for the 2002 Census. As noted in the Census description, "The proportion of principal operators claiming farming or ranching as their primary operation increased significantly since 1997. While there are demographic changes which support this increase, there is a concern that a 2002 forms design change may have also contributed to it." Due to this potential problem, the principal occupation data are not presented here, but it is assumed that days working off the farm can be used in its stead.

Figure A.6 presents the percentage of farmers based on the number of days they reported working off the farm. The majority of farmers report at least some off-farm income. But the divisions between those working off the farm more than two hundred days, or essentially full-time, and those reporting no work off the farm were almost equal—40 and 45 percent, respectively.

Not surprisingly, the distribution of farms within a sales class that report no off-farm work is heavily skewed toward the larger farms. For farms with sales greater than $500,000, 78 percent reported no off-farm work, whereas for farms with sales less than $50,000, only 40 percent reported no off-farm work. Farms in the smallest sales category reported the most full-time employment, while in the largest sales category just 10 percent of the farms reported full-time, off-farm employment.

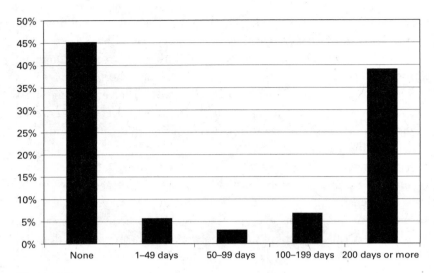

Figure A.6
Percent of U.S. Operators Working Off Farm, 2002

One of the major changes that occurred in the 2002 Census was that data were collected regarding the number of operators per farm. As shown in figure A.7, nearly two-thirds (62 percent) of the farms reported having only one operator. This means more than one-third of U.S. farms can be considered multiple family farms.

The majority of operators, 89 percent, are males (figure A.8). When all operators are included, however, the percent of females increases from 11 to 27 percent. Remember that for most questions, the Census is referring to the principal operator even though some information is gathered on all operators. Figure A.8 also shows the distribution of operators by gender and sales class. Note that over 90 percent of the female principal operators are in the small farm sales category. The percent of principal operators who are female increased by 13 percent. This reflects the increased coverage for the small farms within the Census. Male principal operators decreased by 6 percent.

Direct and Organic Sales

The 2002 Census reported the number of farms along with the levels of direct sales and sales from farms that were certified organic (figure A.9).

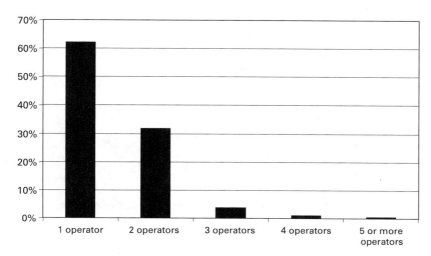

Figure A.7
Percent of U.S. Farms Based on the Number of Operators

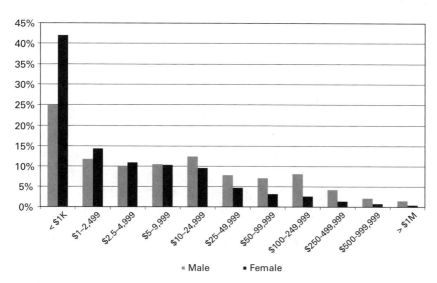

Figure A.8
Percent of U.S. Operators by Gender in Sales Class, 2002

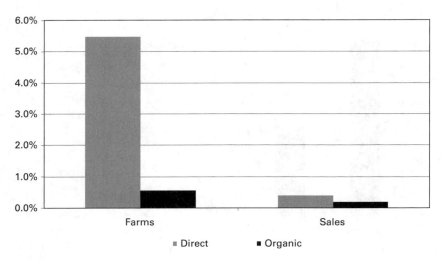

Figure A.9
Percent of U.S. Farms and Sales for Direct and Organic Farms, 2002

This was the first time that questions were asked regarding organic sales. As noted in the Census explanation, "It was the intention of the question to collect only those products that were certified as organic by a government, grower organization, or similar entity."

Direct sales were defined as those "sold directly to individuals for human consumption. This item represents the value of agricultural products produced and sold directly to individuals for human consumption from roadside stands, farmers' markets, pick-your-own sites, etc. It excludes non-edible products such as nursery crops, cut flowers, and wool but includes livestock sales." It is not clear from the Census whether or not direct sales include organic products. Yet it can be assumed that at least some of the organic sales would be direct.

Figure A.10 shows the percent of farms and the percent of sales from farms reporting direct or organic sales; more than 5 percent of the farms reported direct sales and less than 1 percent reported organic sales. The sales in either category represented less than 1 percent of the total agricultural sales in 2002. It is interesting to note that the percentage of direct sales increased by 37 percent from 1997 to 2002. It is not possible to make such a comparison for organic sales.

Only a small percent of farms within any sales category have either direct or organic sales. Figure A.11 depicts the percent of farms within

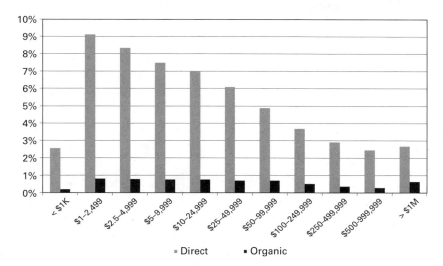

Figure A.10
Percent of U.S. Farms within Sales Class with Direct Sales or Certified Organic, 2002

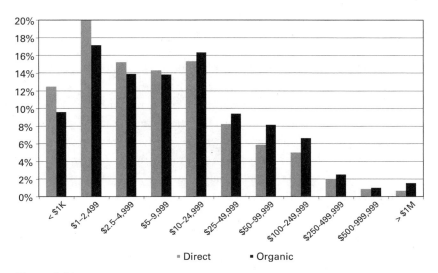

Figure A.11
Distribution of U.S. Farms with Direct Sales or Certified Organic by Sales Class, 2002

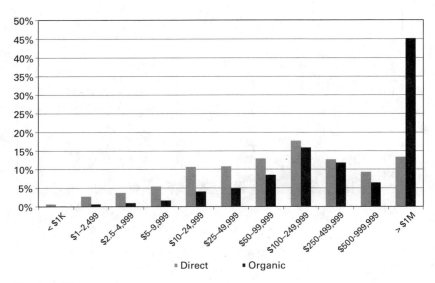

Figure A.12
Distribution of Sales for U.S. Farms with Direct Sales or Certified Organic in Sales Class, 2002

each category with such sales. The smallest category—sales under $50,000—reported the highest percentage of farms (6 percent) with direct sales. The percentage of farms within the sales category reporting organic sales was fairly even across the sales categories.

Figure A.12 portrays the distribution of all farms with direct and organic sales by sales class. The majority of the farms are in the small sales categories. In spite of the high percentage of the farms present in the small sales class, the majority of sales come from the large sales class. Notice that more than half of all organic sales in 2002 were from farms with total sales of $500,000 or more.

Comparing figures A.12 and A.13 illustrates the changes under way in organic agriculture. They show that as organic agriculture becomes more widespread and generally available, the structure of organic agriculture is moving closer to the overall structure of agriculture. It is interesting to note the structure of direct sales as implied in figure A.13. We can see that most—over 40 percent—of the direct sales are made by farms in the middle sales category.

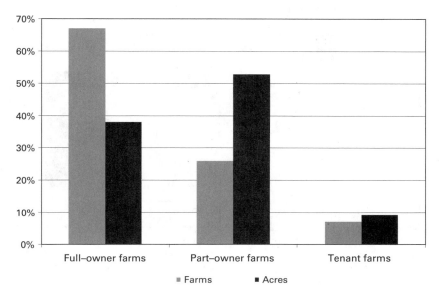

Figure A.13
Percent of U.S. Farms by Ownership Type

Land Tenure

Figure A.13 shows that the majority of U.S. farms are full-owner ones; in other words, the operator owns all the land that is farmed. The majority of acres, however, are farmed by operators who own part but not all of the land they farm. Figure A.14 shows that the percentage of tenant farmers—those who do not own any land—dropped significantly from 1997 to 2002. The land farmed by partial owners is almost equally divided between owned and rented land. Overall, the amount of acres farmed by the partial owners has dropped since 1997.

Figure A.15 portrays the division between owned and rented land in the United States. Just over 60 percent of the land in U.S. farms is farmed by the owner. There was a shift in land tenure between 1997 and 2002. From figure A.16, we can see that the percentage of land owned increased while the percentage of land rented actually decreased 8 percent from 1997 to 2002. Overall, the amount of land in farms decreased approximately 2 percent over the same time period.

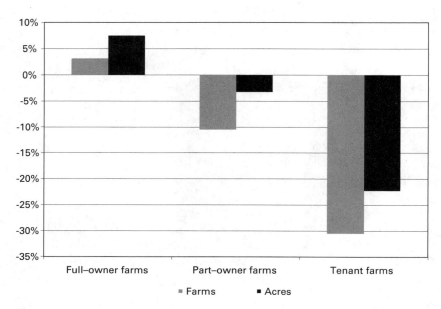

Figure A.14
Percent Change in U.S. Ownership Types, 1997–2002

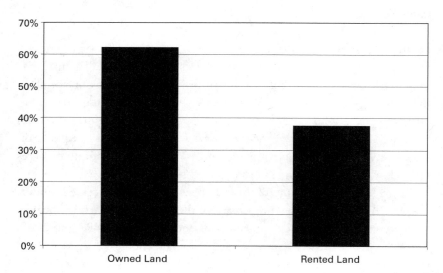

Figure A.15
Owned and Rented Land in the United States, 2002

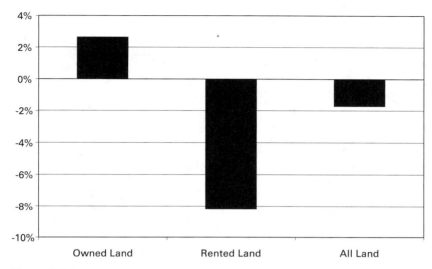

Figure A.16
Percent Change in Owned and Rented Land in the United States, 1997–2002

Full ownership is strongly related to small farms. Almost three-fourths of the small farms are full owners, whereas less than half of the large farms are full owners. The average farm size in 2002 was 441 acres and the median size was 120 acres. Further, just over one-fourth of the land is owned by those farming 10 to 49 acres (figure A.17). This is not surprising given the large number of farms in the low sales categories.

Government Programs

Government programs and payments constitute an increasingly large portion of farm income. Figure A.18 shows the distribution of farms receiving government payments and the total payments by sales class. The distribution depicted in figure A.18 is influenced by several factors. One is that only a third of U.S. farms actually received any direct government payments in 2002. The middle sales category had the highest percentage of farms receiving government payments. For the farms with sales from $50,000 to $499,999, 62 percent received government payments. Another factor for this distribution is that many of the large farms are fruit and vegetable farms. Direct government payments are not available for these commodities.

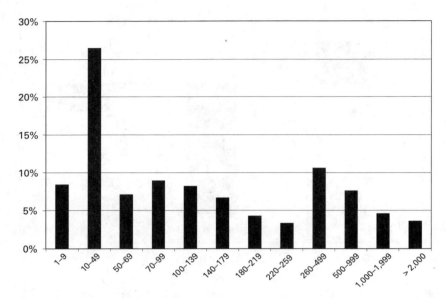

Figure A.17
U.S. Farms by Acre Categories

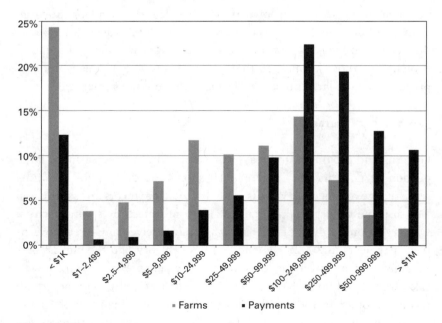

Figure A.18
Distribution of U.S. Farms Receiving Government Payments and Payments by
Sales Class, 2002

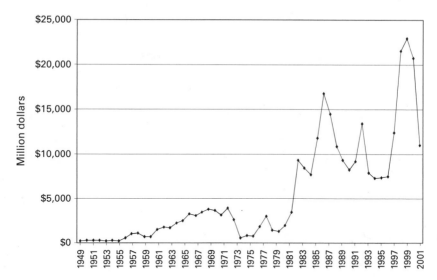

Figure A.19
U.S. Direct Government Payments, 1949–2001

Figure A.19 shows the level of direct government payments to U.S. farms based on data from the USDA's ERS. Notice that there was a tremendous rise in those payments in the mid-1980s. The payment levels have varied since that time, but they still remain at a relatively high level. The level of government payments changes from year to year depending on the yields and prices. There are also changes due to the different features of each farm bill. The 1985 Farm Bill is noted for its major changes in farm policy. These changes included the conservation reserve program and other features that enhanced the level of payments. This major shift is clearly evident in figure A.19.

Figure A.20 shows the government payments as a percent of net farm income. This figure illustrates the change in government involvement in agriculture, especially with respect to income. It can be argued that there is a cost associated with being in the programs so that the percentage in figure A.20 is not entirely accurate. This was especially true prior to the 1996 Farm Bill. Since that time, though, there have not been any set-aside acres, so the cost of involvement in the programs is modest relative to the benefits.

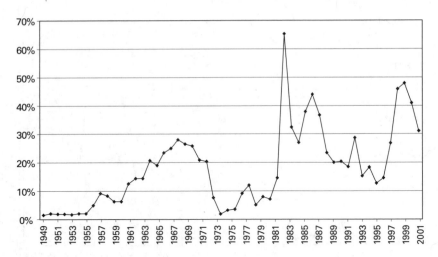

Figure A.20
Government Payments as a Percent of Net U.S. Farm Income, 1949–2001

The government programs exert distributional impacts as well. Figure A.21 captures the distribution of farms that are in the CRP or WRP, both by sales and economic class. The sales classification is based only on sales, and the economic classification is based on sales plus the government payments.

Figure A.22 presents the impact of including government payments on the distribution of the acres in the CRP and the WRP. Similar to figure A.21, when the division is based on the sales class, the majority of the acres are in the smallest class.

Income and Expenses

Figure A.23 shows the change in the total value of agricultural production, total expenses, and net farm income over the past fifty-four years using data from the USDA's ERS. Notice that as the value of production has increased, so has the level of the total expenses. As a result, net farm income has followed a relatively flat trend line over the past few decades.

Figure A.24 presents the same information as figure A.23, but the dollars have been deflated to a constant dollar with 1982–1984 as the base. A slightly different picture emerges when real rather than nominal dol-

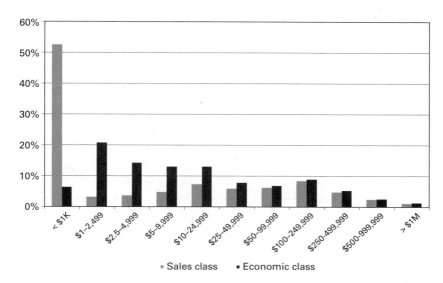

Figure A.21
Distribution of Farms in the CRP and the WRP Based on Sales or Economic
Classification, 2002

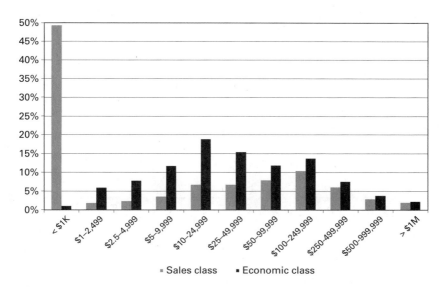

Figure A.22
Distribution of CRP and WRP Acres Based on Sales and Economic Class, 2002

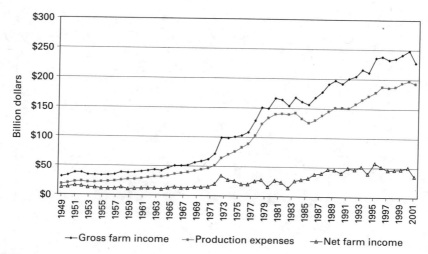

Figure A.23
U.S. Gross and Net Farm Income and Total Expenses, 1949–2001

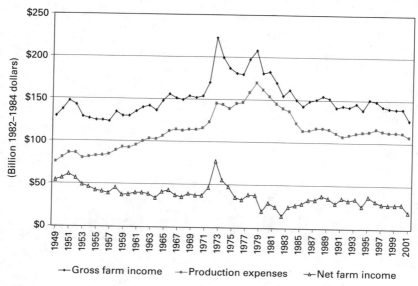

Figure A.24
Gross and Net U.S. Farm Income and Total Expenses, 1949–2001 (1982–1984 Dollars)

Figure A.25
Net as a Percent of Gross U.S. Farm Income, 1949–2001

lars are used. Figure A.24 shows that the value of output, expenses, and net farm income increased through the mid- to late 1970s, but then decreased and have been relatively flat ever since.

Figures A.23 and A.24 reveal the major change that occurred in agriculture during the 1970s and 1980s. In the 1970s, agricultural land values soared. So too did the prices for agricultural commodities. There were at least two major reasons for this dramatic increase. First, there was serious inflation throughout the whole economy. The oil embargo and other factors pushed prices higher. The second reason for the changes in agriculture was the rapid opening and increasing of world trade and exports. There was the Russian wheat deal, the opening of relations with China, and other events that all occurred during the early to mid-1970s. These factors reversed themselves in the early 1980s, as is depicted in figure A.25.

These two figures show that in spite of increasing output, expenses also have risen. This is illustrated in figure A.25, which shows that in the 1950s, net income was approximately 35 percent of sales. In other words, farmers received about $35 of income for every $100 of sales. Over the past decade, net income as a percent of gross has averaged closer to 20 percent. This means a farm would need approximately $100 of sales to generate $20 of income. Figure A.25 helps explain one of the reasons we have seen the expansion in farm size—namely, farms have to be bigger in order to generate the same level of income they did in the past.

Regional Comparisons

Agriculture in the United States varies widely from region to region and even within regions. This section discusses some of the major differences found on a state and regional basis. We examined various areas of the United States, including fourteen different states. Among them were: the Northeast (Vermont, Pennsylvania, and New York), the Southeast (Georgia, North Carolina, and Florida), the Midwest (Michigan, Wisconsin, and Iowa), the Great Plains (Nebraska, Kansas, Texas, and North Dakota), and California.

Sales and Economic Class

Every state followed a pattern similar to the country as a whole, with small farms predominating. There were few differences when measuring size based on sales or economic class. The Midwest states were the only ones with any discernible difference for the small farm category. The other states displayed relatively little difference between the two classification systems regardless of the size. This shows that the government payments are not as important in these areas, relatively, as they are in other states or regions in the country.

Percent Change in Farm Numbers and Sales

The percent change in the number of farms and sales varied somewhat by state. All of the states and the country as a whole showed a decrease in the percent of midsize farms. Five of the fourteen states showed a decrease in the small size as well as the midsize farms. In four of the fourteen states there was an increase in the small and large-size farms. Two of the states showed a decrease in all the categories, and the remaining three states had an increase in the small but not the other two categories.

The variations in patterns with respect to increases or decreases in the small and large categories is due primarily to small numbers of farms in the category. The one common element, regardless of the state or region, was that the midsize farms decreased in numbers. The percentage change in the sales attributed to each sales category was similar across all states and regions. The small and midsize categories had decreased sales while the large category had increased sales.

Percent of Sales and Land

The largest sales category had the highest percentage of sales for most states. The notable exceptions were Wisconsin, North Dakota, Vermont, and Iowa. In these states, the midsize category had the greatest percentage of sales. The pattern of land control by sales class varied somewhat throughout the United States. Most states, however, had a pattern similar to that in figure A.4. California and Florida had the highest percentage of land in the highest sales category. The Northeast states—New York, Pennsylvania, and Vermont—as well as Texas and Georgia showed that more land was controlled by the smallest sales categories. Nebraska and Iowa have more farms in the midsize category than the other states. These two states have almost 40 percent of their farms in the midsize category.

Direct and Organic Sales

Most states showed proportions of direct sales and/or organic farms and sales almost identical to the country as a whole. There were a few notable exceptions, though. States in the Northeast (New York, Pennsylvania, and Vermont) exhibited a considerably higher percentage of farms reporting direct sales. Over 10 percent of the farms in both the smallest and middle sales categories reported direct sales. In California, almost 10 percent of the smallest sales category farms reported direct sales. And in Michigan, approximately 9 percent of both the small and midsize sales categories reported direct sales.

The distribution of farms and sales within a sales class is harder to determine at the state level, especially for the organic categories. The Census will not report information that has the potential of disclosing the individual farm or even approximating whose farm it might be. Because organic—and to a lesser extent direct sales—farms and sales are a small portion of the total sales for many states, the information reported at the state level is somewhat incomplete. Yet for the most part, the individual states followed the pattern established for the entire United States. Most of the organic and direct sales farms fall in the small category.

The highest percentage of organic sales tends to be concentrated in the larger sales groups, in spite of these groups having a lower percentage of the farms. This is one of the issues that occur when dividing by sales; the highest percentage of sales will occur in the higher sales class simply by definition. In the case of organic sales, however, there is an inordinate percentage in the large sales category. For the United States as a whole, 51 percent of the organic sales were in the largest sale category. In California, almost 70 percent of the organic sales were in the largest sales category. Georgia also showed an irregular pattern, with almost 70 percent of the sales occurring in the $50,000 to $500,000 sales category.

Income Patterns

All of the states exhibited a decline in the net income as a percent of gross income category over the 1950 through 2003 period. Nearly all of them started the period in the 30 to 40 percent range, and ended the period in the 15 to 20 percent range. North Carolina, while exhibiting the same general decline as the other states, did begin the 1950 to 1960 period at a considerably higher level. Today, North Carolina is in the same range as the other states, but has fallen much further.

Government Programs

Government programs exert tremendous influence on farm income in the United States. This influence takes several forms and shows definite regional patterns. The distribution of farms receiving government payments and the distribution of those payments reveal a relatively consistent pattern across the states. There are a large percentage of the farms receiving payments in the smallest category and the largest percentages of payments are going to the farms in the middle category.

There are a few exceptions to this pattern. In California, the percentage of payments going to the largest farms is nearly four times the national average, with almost 60 percent of the payments going to this group. In Vermont, and to a lesser extent Nebraska and California, there are more farms in the middle sales category when compared to the national average.

In part the distributional differences can be attributed to the relative importance of the government programs. It is crucial to remember that the government program payments only include the direct payments. They do not include benefits that would accrue through programs such as market orders or allotments.

The impact of the government programs, as measured by the government payments as a percent of net farm income, shows distinct regional patterns. In California, the Southeast states (Florida, Georgia, and North Carolina), and the Northeast states (Pennsylvania, New York, and Vermont), the percent of income coming from the government is much lower than in the other states. The percent of income coming from the government also has been much more constant over time. For example, since 1949 the percent of net farm income received from the government by California and Florida farmers has averaged under or just slightly above 10 percent for the entire period. In contrast, the Midwest and Great Plains states have averaged well over 70 percent in some years.

These results are not surprising given the nature of the farm bill payments and the different agriculture found in different regions of the country. Some regions, such as the Midwest and Great Plains, rely much more heavily on the commodity crops than other areas. California and Florida have a more diverse agriculture including fruits and vegetables, which do not receive direct government payments.

As a result of the changes in the 2002 Farm Bill, most of the states showed a decrease in government support that year. There was a regional exception, however. The states in the Northeast actually showed a fairly significant increase in the percent of the net farm income coming from the government.

Conclusion

The 2002 Census reveals the changing status of the farms and ranches in the middle. These farms once comprised the majority of farms and ranches, but today they are the segment showing the most rapid decline. The United States continues to move rapidly to a farming structure dominated by a large number of small farms and a relatively few large farms.

U.S. government farm programs exert considerable influence on agriculture today. Since 1980, these government program payments have averaged 28 percent of net farm income. Yet in spite of this, and the fact that farms with sales between $50,000 and $500,000 received the most government payments, we are still seeing farms in the middle sales class disappear the fastest.

The farms in the middle control almost half of the land in the United States. As they leave the business, important questions arise regarding who will farm the land and how it will be farmed.

Farms at present face tight margins and struggle to make an adequate income. There are different approaches that can be used to try to generate an adequate income. One approach is to get big. This way attempts to live with the tight margins by increasing the volume. Another approach is to differentiate your product. Organic agriculture is one example of product differentiation. As organic agriculture has become more popular with consumers, we are increasingly seeing the structure for organic agriculture mimic the structure that exists for most of agriculture.

Moving closer to the consumer is another way to live with tight margins. By cutting out the middleperson, the farmer is able to retain a greater amount of the income. The midsize farms appear to be the group taking the most advantage of this approach.

Examining the 2002 Census of Agriculture data reveals several things about the farms in the middle. They are disappearing, and this will in turn change the face of the rural countryside. If the government program payments are intended to maintain this group of farmers then they have failed. Alternative approaches need to be considered.

The results of the 2002 Census leave policymakers with a conundrum. Should policies be geared toward the greatest number of people or the greatest level of production? What policies will be the most effective, and for what goals? If land use is the issue, then it appears that the midsize farms—those with sales between $50,000 and $500,000—should be the target. Yet if the greatest number of people is the target, then programs should be geared toward the smallest farms. But if production is the key issue, then targeting only the largest farms would be the most effective. Trying to create and administer a policy geared toward all farming levels will not succeed because of the vast differences between the farms.

The process of compiling the data for the 2002 Census was changed to more accurately reflect the number of small farms and proved to be successful in capturing information from more of these farms. Unfortunately, at the same time that more of these farms were being enumerated, their inclusion in the data set masked much of what was happening in farming. The reported drop in farm numbers would have been almost four times as great had the smallest farms been excluded.

Policymakers need to have adequate information if they are to make the best policy decisions. People need to be aware of what is happening in agriculture. Only with an adequate Census and realistic summaries will that information be available.

Contributors

Elizabeth Barham is Assistant Professor of Rural Sociology at the University of Missouri–Columbia. She is coeditor with Bertil Sylvander of the forthcoming book *Geographical Indications for Food: Local Development and Global Recognition* (2007).

Sandra S. Batie is the Elton R. Smith Professor in Food and Agricultural Policy in the Department of Agricultural Economics at Michigan State University. Sandra chaired the 1993 National Research Council (National Academy of Science) committee that produced the book *Soil and Water Quality: An Agenda for Agriculture* (1993).

Eileen Brady is co-owner of New Seasons Market, a neighborhood grocery chain in Portland, Oregon, that offers an extensive selection of regionally grown and processed foods. She founded the Food and Farms program at Ecotrust, a nonprofit providing conservation and economic development solutions to communities from Alaska to California.

Frederick Buttel was Professor of Sociology and Environmental Studies at the University of Wisconsin–Madison until his death in 2005. He was the author, editor, or coeditor of a number of books, including *Governing Environmental Flows: Global Challenges to Social Thought* (coedited with Gert Spaargaren and Arthur P. J. Mol and published by MIT Press in 2006).

Peter Carstensen is the George H. Young-Bascom Professor of Law at the University of Wisconsin at Madison. He is the coeditor of Federal Statutory Exemptions from Antitrust Law (Monograph 24, ABA Section of Antitrust, 2007).

Kenneth A. Dahlberg is Professor Emeritus of Political Science and Environmental Studies, Western Michigan University. He is the author of *Beyond the Green Revolution* (1979), the editor of *New Directions for Agriculture and Agricultural Research* (1986), and coeditor of *Environment and the Global Arena* (1985) and *Natural Resources and People* (1986).

Mike Duffy is a Professor of Economics at Iowa State University in Ames, Iowa. He is also the Director of the Beginning Farmer Center and the Chair of the Graduate Program for Sustainable Agriculture at Iowa State.

Thomas W. Gray is a Rural Sociologist in the Law, Policy, and Governance area of the USDA's Rural Development–Cooperative Programs in Washington, DC.

Shelly Grow is a Conservation Biologist for the Association of Zoos & Aquariums in Silver Spring, Maryland.

Amy Guptill is an Assistant Professor of Sociology at SUNY College at Brockport in Brockport, New York. She is the coauthor of *Food and Society: Principles and Paradoxes*, forthcoming.

William Heffernan is Professor Emeritus of Rural Sociology at the University of Missouri–Columbia. He serves on the board of the Ozark Mountain Pork Cooperative, an entity producing sustainably and humanely raised pork.

Mary Hendrickson is an Extension Associate Professor of Rural Sociology at the University of Missouri–Columbia, where she directs MU Extension's Food Circles Networking Project.

Elizabeth Higgins is the Issue Leader for Family and Consumer Science programs at Cornell Cooperative Extension of Ulster County, New York.

Fred Kirschenmann is Distinguished Fellow for the Leopold Center for Sustainable Agriculture at Iowa State University.

David Lind is Assistant Professor of Sociology at Goshen College in Goshen, Indiana.

Thomas A. Lyson was Liberty Hyde Bailey Professor of Development Sociology at Cornell University until his death in 2007. He was the author of *Civic Agriculture: Reconnecting Farm, Food, and Community* (2004).

Caitlin O'Brady is a Field Guide with the Moab Canyonlands Institute in Utah.

Rich Pirog is the Associate Director and Marketing and Food Systems program leader at the Leopold Center for Sustainable Agriculture at Iowa State University.

Daryll E. Ray is a Professor of Agricultural Economics. He is Director of the Agricultural Policy Analysis Center and holds the Blasingame Chair of Excellence at the University of Tennessee–Knoxville. He developed POLYSYS, a computer-based analytical system that models the impact of policy changes on the U.S. agricultural sector.

Harwood D. Schaffer is a Research Associate at the Agricultural Policy Analysis Center at the University of Tennessee–Knoxville.

G. W. Stevenson is Senior Scientist with the Center for Integrated Agricultural Systems at the University of Wisconsin–Madison.

Rick Welsh is an Associate Professor of Sociology at Clarkson University in Potsdam, New York.

Index